Advanced fiber spinning technology

Edited by
Professor T Nakajima
President, Society of Fiber Science & Technology, Japan

English edition edited by
K Kajiwara and J E McIntyre

WOODHEAD PUBLISHING LIMITED
Oxford Cambridge New Delhi

Published by Woodhead Publishing Limited, Abington Hall, Granta Park,
Great Abington, Cambridge CB21 6AH, UK
www.woodheadpublishing.com

Woodhead Publishing India Private Limited, G-2, Vardaan House,
7/28 Ansari Road, Daryaganj, New Delhi – 110002, India

English edition, first published 1994, Woodhead Publishing Limited
Reprinted 1996, 2000, 2007, 2009

Japanese edition, first published 1992, by Kobunshi Kankokai, Kyoto, Japan

British Library Cataloguing in Publication Data
A catalogue record for this book is available from the British Library.

ISBN 978-1-85573-182-0

Contents

Contributors

S R Allen (Senior Research Associate, Advanced Fibers Systems Business, Du Pont, Richmond, Virginia, USA)
He received his PhD in Polymer Science and Engineering at the University of Massachusetts. He is the author of numerous polymer/fiber-related publications and patents and was awarded the Fiber Society Award for Distinguished Achievement in 1992.

Kaoru Ban (Senior Coordinator, Fibers and Textiles Technical Center, Toyobo Research Institute, Toyobo Co, Ltd, Katada, Ohtsu-shi, Shiga-ken, Japan)
He graduated from Kyoto Institute of Technology, and joined Toyobo Co, Ltd, in 1969.

Ken-ichi Katayama (President, Takuma National College of Technology, Takuma-cho, Mitoyo-gun, Kagawa-ken, 769-11 Japan)
He graduated from Kyoto University and joined Osaka Municipal Technical Research Institute as a research worker in 1950. He was a senior research worker at Asahi Chemical Industry Co. Ltd. from 1952 until 1972, and became an associate professor at Institute for Chemical Research, Kyoto University in 1972. He was promoted to a full professor in 1978, and then to a president of Takuma National College of Technology in 1991.

Ei-ichi Kubo (General Manager of Non-Woven Fabrics Department, Unitika Ltd, Uji-shi, Kozakura 23, 611 Japan)
He graduated from Kyoto Institute of Technology and joined Unitika Ltd in 1964. He is engaged in research and development of spinning technology and non-woven fabrics. He was awarded the Technology Prize from the Society of Fiber Science and Technology, Japan, in 1993, for his achievement in developing new non-woven fabrics.

Masao Matsui (Retired Researcher, Kanebo Ltd, Osaka, Japan)
He graduated from Kyoto University and joined Kanebo Ltd in 1957. He was involved in developing new synthetic fibers, bi-component conjugate fibers, aesthetic fibers, and ultra-fine fibers at Kanebo Ltd. He was a general manager of the Research Center for Synthetic Fibers of Kanebo Ltd, and was invited to Kyoto Institute of Technology as a guest professor from 1992 until 1994. He was awarded Dr. Eng. from Kanazawa University in 1994.

Yasuhiro Murase (Director and Assistant General Manager of Fibers Technology and Production Division, Teijin Ltd)
He graduated from Osaka University and joined Teijin Ltd in 1958. He was engaged in process designing, construction and operation of polyester staple fiber plant and the research and development of new speciality polyester staple fibers (1958–81), and research into high performance or high functional fibers at Teijin Fiber & Textile Research Laboratories (1981–89). He was promoted to General Manager of Teijin Fiber & Textile Research Laboratories in 1989 and then to his present position in 1994.

Akihiko Nagai (Associate Professor, Seitoku University, Junior College, Department of Life and Culture, Tokyo, Japan)
He graduated from Kyushu University and joined Teijin Ltd in 1968. He has been engaged in research and development of fibers for industrial usage at the Matsuyama and Iwakuni Plants (1968–81), and fundamental research into high performance fiber at Teijin Fiber and Textile Research Laboratories (1981–93). He moved to his present position in 1993.

Junyou Nakagawa (Chief Researcher, Technical Research Center, Kuraray Co, Ltd, Kurashiki, Okayama-ken, Japan)
He graduated from the Graduate School of Hokkaido University (M. Eng.), and joined Kuraray Co, Ltd, in 1968. He was awarded the Technology Prize of Society of Fiber Science and Technology, Japan, in 1983. He has been engaged in the development and application of fibers made from thermotropic liquid crystal polymers since 1985.

Yasuo Ohta (Assistant Research Coordinator, Fiber Research Laboratory, Toyobo Research Institute, Toyobo Co, Ltd, Katada, Ohtsu-shi, Shiga-ken, Japan)
He graduated from Kyoto University and joined Toyobo Co, Ltd, in 1984.

Miyoshi Okamoto (Managing Director, Toray Industries Inc, Ohtsu-shi, Sonoyama 3-2-2, 520 Japan)
He graduated Nagoya Institute of Technology in 1960, and was awarded

Dr. Eng. in 1989. He joined Toray in 1960 and has been Manager of Ecsaine Research Laboratory, Manager of Speciality Fabrics Research Laboratory, Manager of New Fiber Research Laboratory and General Manager of Research Laboratories for Fibers and Textiles. He became a Director of Toray Industries Inc. in 1989 and a Managing Director in 1992. He has received numerous awards including the Carothers Memorial Medal from the Textile Institute (1993), and in 1989 Toray did him the unique honour of founding the Okamoto Research Laboratory to give top researchers the freedom to create the technology of the next generation.

Katsuhiko Shimada (Manager, Toyama Plastics Plant, Mitsubishi Rayon Co, Ltd, Toyama-ken, Japan)
He graduated from Osaka City University, received an MSc and joined Mitsubishi Rayon Co, Ltd, in 1982. He has been engaged in development of transparent polymers for POF.

Masaki Tsuji (Associate Professor, Institute for Chemical Research, Kyoto University, Uji, Kyoto-fu, 611 Japan)
He graduated from Kyoto University in 1975, and was awarded Dr. Eng. in 1981. He became a Postdoctorial Research Fellow at McGill University in 1981, and returned to the Institute for Chemical Research, Kyoto University in 1994. He was promoted to associate professor in 1994.

Takashi Tsurumi (Manager, Cashmilon Plant, Asahi Chemical Industry Co, Ltd, Samejima 2-1, Fuji-shi, 416 Japan)
He graduated from Tokyo University and joined Asahi Chemical Co Ltd, in 1966. He was awarded Dr. Eng. from Kyushu University in 1991, and was promoted to his present position in 1993.

Masaharu Watanabe (Plant Manager of Okazaki Plant, Unitika Ltd, Okazaki-shi, Aichi-ken, Japan)
He graduated from Kyushu University and joined the research and development center of Unitika Ltd at Uji in 1963. He is engaged in the research and development of spinning technology especially in the field of high-speed spinning and non-woven fabrics. He was promoted to his present position in 1992.

H H Yang (Retired Research Fellow, Advanced Fibers Systems Business, Du Pont, Richmond, Virginia USA)
He earned his PhD in chemical engineering at the University of Michigan. He is the author of three books on high performance fibers, is an honorary professor at the China Textile University in Shanghai, People's Republic of China, and holds numerous polymer-related patents.

Hiroshi Yasuda (General Manager, Fiber Research Laboratory, Toyobo Research Institute, Toyobo Co, Ltd, Katada, Ohtsu-shi, Shiga-ken, Japan)
He graduated from Kyushu University and joined Toyobo Co, Ltd, in 1964.

Preface

This book covers the new developments of fiber spinning science and technology. Much of the success of Shingosen, the newly specified synthetic fiber which has been developed in Japan, depends upon a skilful application of the spinning technology.

This book gives a comprehensive presentation of the techniques with the theoretical background for an understanding of them.

I hope that this volume will be found useful for readers in many countries and will contribute to a better understanding of the science and technology of the field of spinning.

Toshinari Nakajima
President of The Society of Fiber Science and Technology, Japan

Introduction

Some of the most interesting developments of the last few decades in the field of fiber manufacture have been the result of intensive study in Japanese industry and research institutes. Features of such research are a determination to push the technology towards its limits and a willingness to embark rapidly on commercial exploitation of the results. An unusual combination of intensive competition and co-operation between the many substantial companies involved results in a rapidly evolving technology. Consequently the rate of introduction of new processes and products is considerably higher in Japan than in other countries.

This book was originally published in Japanese by the Society of Fiber Science and Technology, Japan, in order to present a thorough scientific and technological review of advances in fiber production, and is now published in English to mark the 50th anniversary of the foundation of the Society. The emphasis is upon developments either originated or adopted in Japan. With the exception of a chapter about aramid fibers contributed from Du Pont, the authors are leading Japanese academic and industrial research workers.

In editing the translation we have tried to make it as friendly to English-speaking readers as possible. We have, however, retained many references to journals and patents written in Japanese, since it has seldom been possible to identify clearly equivalent material in other languages.

K Kajiwara
Kyoto Institute of Technology
J E McIntyre
The University of Leeds

Japanese Patents: explanatory note

Japanese patents appear in two stages, with different numbers at each stage. The numbers consist of the year followed by a number within that year. The years are numbered from the beginning of each Imperial era. Showa is from 25 December 1926 and Heisei from 7 January 1989.

In this translation the two stages are termed 'laid open' and 'examined', respectively. The years have been altered to the year AD, which in nearly all cases can be converted into the Japanese year by subtracting 25 from years up to '88 and 88 from years '89 onwards.

1

Fundamentals of spinning

Ken-ichi Katayama
Takuma National College of Technology, Takuma-cho, Japan
Masaki Tsuji
Institute for Chemical Research, Kyoto University, Japan

There are a great many subjects covered by the heading 'Fundamentals of spinning'. In this chapter, however, we confine ourselves to a description of the fundamentals of mathematical simulation for spinning and of new findings on structural formation during spinning and fiber structure. This description should be undertaken for each of the three types of spinning: melt spinning, dry spinning and wet spinning. However, we are concerned here mainly with melt spinning because it is the easiest for us to formulate and accordingly its theory is the most sophisticated of the three. The others are described only briefly with literature references to the details. Nevertheless, the authors hope that this short treatise will help the readers to understand other parts of this book.

1.1 Introduction

The viscose rayon method was developed towards the end of the last century and the melt spinning method for synthetic fibers was established in the early part of the 1930s. In the beginning of the history of spinning, progress in spinning technique was mainly made by accumulating empirical facts; that is to say, by repeating a set of procedures such as setting a spinning condition and measuring the resultant properties and structures of the spun fibers. There were few studies on physiochemical changes and on structural formation in the spun fibers between the spinneret and the take-up device. With the rapid advance of the synthetic fiber industry in the 1940s, a strong need arose to understand the basics of the spinning process in order to improve the productivity and quality control of the fibers. Consequently, towards the end of the 1950s, Ziabicki published a series of papers concerning melt spinning in which the spinning process was analyzed mathematically as an engineering problem: the papers served as a powerful incentive to researchers in this field of study. About the middle of the 1960s, Kase and Matsuo[1,2] established a method for the quantitative description of the melt spinning process based on hydrodynamics, rheology and the theory of thermal conduction. Subsequently, Katayama *et al*[3] studied structural formation and crystallization during a melt spinning process by using a special model spinning apparatus. From then on, studies in this field have been extensively carried out in Japan and elsewhere and most of the results are to be found in several references.[4–9]

In melt spinning, we can predict the diameter and temperature and the tension in a running filament if the spinning conditions and the rheological properties of a polymer used in the spinning process are given; the predicted values are, of course, in good agreement with the experimental results. Such a prediction, however, can be made only when no significant crystallization occurs during the spinning process. If crystallization must be taken into consideration, it becomes increasingly difficult for us to carry out mathematical simulation of the quantities mentioned above.

In this case, full quantitative knowledge of the following four points is needed for performing a mathematical simulation of melt spinning:

1 Molecular orientation caused by elongational melt flow.
2 Influence of molecular orientation on crystallization kinetics.
3 Changes in the rheological properties of the polymer caused by molecular orientation and crystallization.
4 Kinetics of non-isothermal crystallization.

The correlation between the above points is shown in Fig. 1.1. In this figure, an arrow indicates that an item from which the arrow starts influences another item to which the arrow is pointing.

In the theory of Kase and Matsuo, mean values of temperature and of stress over the whole area of a transverse section of the filament were used. In the process of high speed spinning, however, the variables such as temperature, stress, orientation and crystallinity must be expressed as functions of the radial distance from the central axis of the filament as well as the distance from the spinneret. For example, consideration of the radial distribution of these variables is inevitable in discussing inhomogenous structures such as skin–core structure. In order to understand the spinning process, it is indispensable for us to know how structure will be formed during the process as well as to carry out the detailed technological analysis of the process. In the following part of this chapter, the fundamental equations describing the melt spinning process accompanied by crystallization will be developed. After this the oriented crystallization, which is closely related to the structure formation during the spinning process, will be discussed and the differences between oriented and unoriented crystallization will be highlighted. Then the importance of gelation in the process of solution spinning and of phase separation will be mentioned. Finally, examples of mathematical simulation of high speed melt spinning will be demonstrated.

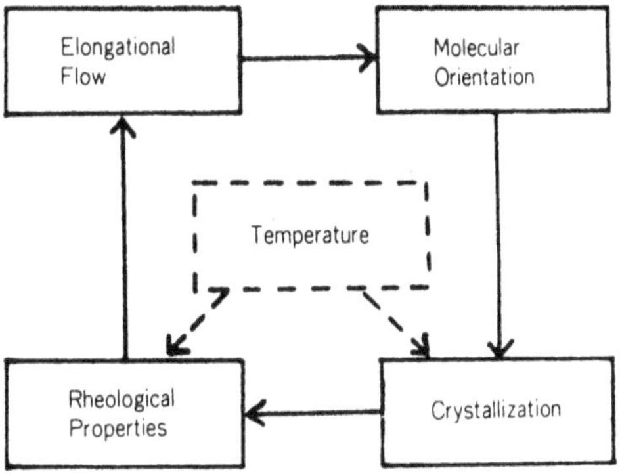

1.1 Correlation between factors governing the melt spinning process

1.2 Technological analysis of spinning

For the technological analysis of spinning, three fundamental equations are derived from the conversion of energy, the conversion of momentum and the conversion of matter, respectively. Here, for simplification, the two differential operators are defined as follows:

$$\nabla = \mathbf{i}\frac{\partial}{\partial x} + \mathbf{j}\frac{\partial}{\partial y} + \mathbf{k}\frac{\partial}{\partial z} \tag{1.1}$$

$$\frac{\mathrm{D}}{\mathrm{D}t} = \frac{\partial}{\partial t} + V_x\frac{\partial}{\partial x} + V_y\frac{\partial}{\partial y} + V_z\frac{\partial}{\partial z} \tag{1.2}$$

where t is the time, (x,y,z) the spatial co-ordinates, $\mathbf{V}(V_x,V_y,V_z)$ the velocity of polymer, and **i**, **j** and **k** are the unit vectors in the x, y, and z directions, respectively.

1.2.1 Fundamental equations (melt spinning)[10]

1.2.1.1 Conversion of energy

Consider an arbitrary small-volume element $\mathrm{d}v$ fixed in space along the spinning path (Fig. 1.2) and an enthalpy change in the element. The fundamental equation for conservation of energy can then be derived.

1.2.1.2 Inflow of heat through conduction

We assume that the thermal conductivity of the polymer K_c, is independent of temperature T. The net inflow of heat in unit time conducted through the xy-plane into the element, the centre of which is located at (x,y,z), is expressed as

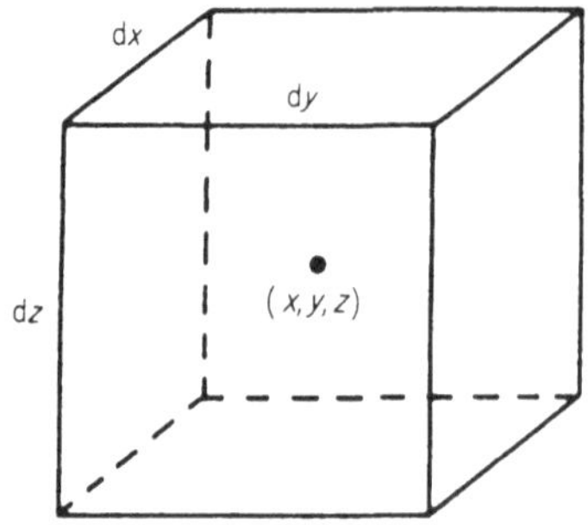

1.2 Volume element

$$\left(K_c \frac{\partial T}{\partial z}\bigg|_{z+\mathrm{d}z/2} - K_c \frac{\partial T}{\partial z}\bigg|_{z-\mathrm{d}z/2}\right)\mathrm{d}x\mathrm{d}y = K_c \frac{\partial^2 T}{\partial x^2}\,\mathrm{d}x\mathrm{d}y\mathrm{d}z \quad [1.3]$$

Accordingly, by summing such quantities for all the three directions, heat transferred into the element in unit time is

$$K_c \nabla^2 T \mathrm{d}v \ (\nabla^2\text{: Laplacian}) \quad [1.4]$$

1.2.1.3 Inflow of heat accompanying transfer of matter

For the polymer, the specific heat at constant pressure, the heat of crystallization, the crystallinity and the density are expressed as C_p, ΔH, X and ρ respectively. We assume that C_p, and ΔH are independent of T and that the enthalpy per unit mass (H) is a function of T and X.

The net enthalpy in unit time which flows into the volume element dv through the xy plane is

$$(-\rho H V_z|_{z+\mathrm{d}z/2} + \rho H V_z|_{z-\mathrm{d}z/2})\mathrm{d}x\mathrm{d}y = -\frac{\partial(\rho H V_z)}{\delta z}\,\mathrm{d}v \quad [1.5]$$

Accordingly, by summation of such quantities for all the three directions,

$$-\nabla . (\rho H \mathbf{V})\mathrm{d}v \quad [1.6]$$

Using the equation of continuity $\partial\rho/\partial t = -\nabla . (\rho \mathbf{V})$ (modified equation 1.4),

$$-\nabla . (\rho H \mathbf{V})\mathrm{d}v = H \frac{\partial p}{\partial t} - \rho \mathbf{V}.\nabla H \quad [1.7]$$

1.2.1.4 Conservation of enthalpy

Here, we direct our attention to the conservation of enthalpy within dv

$$\frac{\partial(\rho H)}{\partial t} = K_c \nabla^2 T + H \frac{\partial \rho}{\partial t} - \rho \mathbf{V}.\nabla H \quad [1.8]$$

Since $\partial H/\partial T = C_p$ and $\partial H/\partial X = -\Delta H$,

$$\nabla H = C_p \nabla T - \Delta H . \nabla X \qquad [1.9]$$

Substitution of this relation into equation 1.8 gives

$$\rho \frac{\partial H}{\partial t} = K_c \nabla^2 T - C_p \rho V . \nabla T - \rho \Delta H \mathbf{V} . \nabla X \qquad [1.10]$$

For the steady state, equation 1.10 becomes

$$\mathbf{V} . \nabla T - \kappa \nabla^2 T = \frac{\Delta H}{C_p} \mathbf{V} . \nabla X \qquad [1.11]$$

where $\kappa = K_c / (\rho C_p)$ is the thermal diffusivity. Generally speaking, $\mathrm{D}/\mathrm{D}t$ can be regarded as $\mathbf{V}.\nabla$.

1.2.1.5 Conservation of matter

The following equation is derived on the basis of the balance of the matter which flows into the volume element as shown in Fig. 1.2.

$$\frac{\mathrm{D}\rho}{\mathrm{D}t} = - \rho \nabla . \mathbf{V} \qquad [1.12]$$

Assuming that the polymer is not a compressible material,

$$\nabla . \mathbf{V} = 0 \qquad [1.13]$$

1.2.1.6 Conservation of momentum

Based on the balance of the momentum which flows into the volume element (Fig. 1.2). the following equation is obtained

$$\rho \frac{\mathrm{D}\mathbf{V}}{\mathrm{D}t} = - \nabla P - [\nabla . \mathbf{p}] + \rho \mathbf{g} \qquad [1.14]$$

where P is the pressure, in a normal sense, of isotropic fluid, $\mathbf{p}$ the excess stress tensor and $\mathbf{g}$ the acceleration of gravity. $[\nabla.\mathbf{p}]$ is a vector, and, for example, one of its components is expressed as

$$[\nabla . \mathbf{p}]_x = \frac{\partial p_{xx}}{\partial x} + \frac{\partial p_{xy}}{\partial y} + \frac{\partial p_{xz}}{\partial z} \qquad [1.15]$$

Under the assumption of non-compressibility, P has no physical meaning, but will be determined by the boundary condition. When the spinning direction is chosen as the z-axis, the quantity, $p_{zz} - p_{xx}$, corresponds to the tensile stress.

1.2.1.7 Other equations

In addition to the three equations corresponding to conservation of energy, matter and momentum, we must take account of constitutive equations (rheological equations[8]), equations of crystallization kinetics, the equation concerning molecular orientation (birefringence Δn) and the thermodynamic equation of state. Before turning to the discussion of each of these equations, we shall direct our attention to the number of unknown variables and of equations. Table 1.1 shows the number of unknown variables and of equations, both of which total 14. In principle, therefore, we can solve the equations if the boundary conditions are given. The important equations governing the boundary conditions are the equation of thermal conduction on the surface of filament and that of air resistance.

Table 1.1 Numbers of unknown variables and equations

Unknown variables	
V_x, V_y, V_z	3
p_{xx}, p_{yy}, p_{zz}, p_{yz}, p_{zx}, p_{xy}	6
P, T, ρ	3
X, Δn	2
	Total 14
Equations	
Equation for conservation of energy	1
Equation of conservation of matter	1
Equations for conservation of momentum	3
Constitutive equations	6
Equation of crystallization kinetics	1
Equation concerning molecular orientation	1
Equation of state	1
	Total 14

1.2.1.8 Constitutive equations

Various complicated formulae have been proposed so far as constitutive equations which relate the excess stress tensor **p** to the thermal and deformation history.[8] The simplest example is Newton's equation of viscosity.

1.2.1.9 Equation of crystallization kinetics

The nucleation rate of polymers at a constant temperature is greatly accelerated by molecular orientation. For the present, however, there is no general formula expressing the quantitative relation between the crystallization rate and molecular orientation. Furthermore, the crystallization under molecular orientation may be different from ordinary unoriented crystallization, which is expressed in terms of the nucleation rate, the growth rate of the nucleus and the mode of geometrical growth. At any rate, the process of structural change in oriented crystallization has never been clarified. This will be discussed in a subsequent part of this chapter. If crystallization kinetics are described in the form of the Avrami equation, $X = 1 - \exp(-K_A t^n)$, with increasing molecular orientation the rate constant K_A increases rapidly and the Avrami index n decreases to unity[11,12] or even below.[13] In reality, the Avrami equation applies only in the early stages of crystallization. In addition, it should be noted that secondary crystallization becomes prominent in the advanced stages of crystallization.

Adopting the birefringence Δn (or the tensile stress σ) as a parameter relating to molecular orientation, we can tentatively express the rate constant of crystallization K_A at a constant temperature as a function of T and Δn. Nevertheless, there still remains a question of how to utilize the corresponding data for describing non-isothermal crystallization. Though we have no approved answers to this question, the following expression for the crystallinity X can be adopted without any gross errors[14,15]

$$X = 1 - \exp\{-(\int_0^t K(T,\Delta n)\mathrm{d}t')^n \} \qquad [1.16]$$

where the equation for kinetics of isothermal crystallization is expanded into the case of non-isothermal crystallization and the relation, $K(T, \Delta n) = \{K_A (T, \Delta_n)\}^{1/n}$, is assumed.[14,15] Equation 1.16 is an integral equation which describes X using the history of T and Δn (namely, σ).

1.2.1.10 Relation between tensile stress and birefringence

An experimental linear relation between the birefringence Δn and the tensile stress σ during spinning has been reported for small σ ($\sigma < 3 \times 10^7$ dyne/cm^2 for polyethylene terephthalate [PET]).[16] This is easily understood because the theory of rubber elasticity[17] has definitely proved that σ/KT (K is Boltzmann's constant) is directly proportional to Δn for small σ (Gaussian chain approximation). In the case of high speed spinning, however, we must take account of the relation between Δn and σ for large σ because σ readily reaches up to 10^8 – 10^9 dyne/cm^2. While σ can approach infinity, Δn has its maximum value defined by the intrinsic birefringence Δn_{in}. Hence, with increasing σ, the rate of increase of Δn decreases and Δn itself approaches a constant value. Ziabicki and Jarecki[18] obtained a numerical relation between the value of Δn in the steady state (Δn_{st}) and σ/KT for Langevin chains (Fig. 1.3).

The above applies to steady-state flow but the spinning process is in the non-steady state in terms of molecular orientation. Thus, Δn is always smaller than Δn_{st}, which is the value of Δn in the steady state for the present σ. Assuming a single delay time τ,

$$\mathrm{D}\,\Delta n/\mathrm{D}t = (\Delta n_{st} - \Delta n) / \tau \qquad [1.17]$$

then through σ/KT, Δn_{st} is a function (and through T, τ is a function) of position and time.

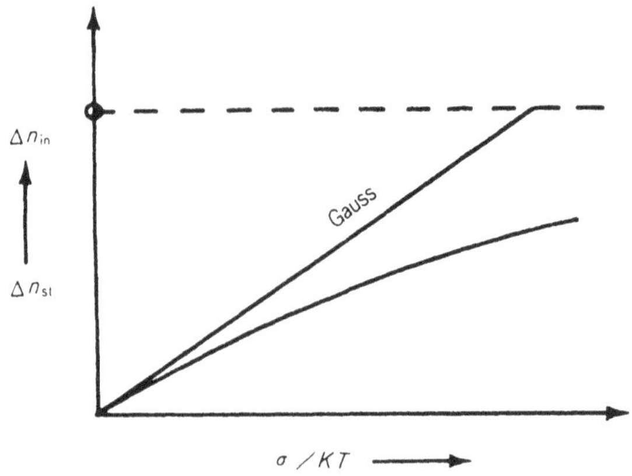

1.3 Relation between the birefringence in the steady state and σ/KT.[18]

1.2.2 Solution spinning

1.2.2.1 Dry spinning

In the process of dry spinning, the fiber structure is formed by forcing the polymer solution out through a fine nozzle and then evaporating the solvent. Accordingly, dry spinning can be treated as a problem of structure formation in a two-component system of the polymer and its solvent. Technological analysis of the dry spinning is, therefore, much more difficult than that of melt spinning which is treated as a problem of structural formation in a one-component system. An equation of diffusion for a two-component system is needed to describe structure formation within the filament, and an equation of evaporation rate of the solvent at the boundary between two phases is also needed on the surface of the filament. Moreover, some equations used in the mathematical treatment of melt spinning should be modified to apply them to dry spinning; details are found in the references.[4,19,20]

1.2.2.2 Wet spinning[5]

In the process of wet spinning, where a solution of the polymer is forced out through a nozzle into a non-solvent for the polymer, mass transfer for both the solvent and non-solvent must be considered. Technological analysis of wet spinning is, accordingly, much more complicated than that of dry spinning. In a spinning process accompanied by chemical reactions, quantitative analysis is almost impossible. As for the spinning process without any chemical reactions, the phase diagram using triangular co-ordinates for a three-component system of polymer (P), solvent (S) and non-solvent (N) can be drawn. In this case, it is the ratio F_N/F_S of the flux of solvent from the filament into a spinning bath (F_S) to that of non-solvent from the bath into a filament (F_N) that determines which path is followed on the phase diagram in the process of fiber formation.

1.3 Structural formation during spinning

The structural changes in the spinning process, crystallization, gelation, and phase separation, are discussed here.

Crystallization arising from the state of molecular orientation in a polymer solution, melt or amorphous solid is termed 'oriented crystallization'. Crystallization during spinning is a typical example of oriented crystallization. Structure formation in oriented crystallization is of great interest because it is a phenomenon reflecting the nature of the macromolecule.

Since the 1960s, studies on oriented crystallization have been carried out extensively and the publications so far are too many to mention. Of all these studies, that of flow-induced crystallization of polyethylene in particular attracted researchers' attention. When polyethylene is crystallized from solution (concentration 0.5–5%) by stirring the solution, what is called the shish-kebab structure[21,22] is formed.

The following are general features of oriented crystallization:

1 The morphology of the crystallized materials changes according to the degree of molecular orientation.
2 With increasing degree of molecular orientation, the temperature which gives the maximum rate of crystallization goes up and, in some cases, the maximum rate itself also increases by several orders of magnitude.
3 The mechanism of oriented crystallization may be very different from that of non-oriented crystallization.

1.3.1 Oriented crystallization from melts

The knowledge that we have so far acquired of structure formation from oriented melts of flexible polymers is summarized as follows:

1 In advance of crystallization, a spatial non-uniformity of density occurs.[23] A domain with higher density has a rod-like shape elongated along the direction of molecular orientation. The diameter of the domain depends strongly on the degree of molecular orientation and decreases with increasing degree of orientation. The domain, generally speaking, is of a size which can be seen with a light microscope.
2 Within such a domain, density fluctuations on a scale of several tens of nm occur within a stacked lamella-like structure in which constituent 'lamallae' are developed perpendicularly to the direction of molecular orientation. The fluctuation can be detected by meridional small-angle

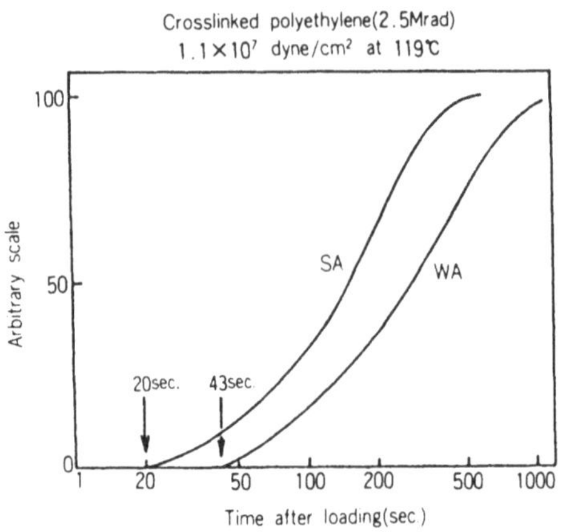

1.4 Oriented crystallization of crosslinked polyethylene. When wide-angle (WA) and small-angle (SA) X-ray scatterings are observed simultaneously in the course of crystallization at a constant load, SA scatterings appear first before WA scatterings, which indicates that the density fluctuation is evolved in the form of lamellae elongated perpendicularly to the stretching direction.

X-ray scattering.[3] At this stage of structure formation, wide-angle X-ray diffraction from the crystalline state is still not observed. In conclusion, there exists a mesomorphic state during transformation from the amorphous state to the crystalline (see Fig. 1.4 and Reference 3).

3 In the shish-kebab structure and the row structure (a structure showing an appearance like a Japanese mat, 'tatami'), crystalline lamellae are stacked in the direction of molecular orientation. Such morphology was interpreted by Keller and co-workers[24,25] as a structure consisting of folded-chain lamellar crystals grown epitaxially on a long and slender nucleus composed of extended chains. Row structure, however, can readily emerge even from very weakly-oriented melts, and in a thin film of the polymer we can observe many nuclei which are aligned in a line in the rod-like domain.[26] It is, therefore, presumed that the concept of a nucleus composed of extended chains would be a product of too extreme modeling. We should rather venture to say that the characteristic of structure formation in oriented crystallization is that oriented nuclei are apt to align themselves in a line. In thin films crystallized under molecular orientation with a rather high degree of orientation, however, a crystalline domain of about 200 nm in length and about 15 nm in width was identified by high-resolution electron microscopy as a domain in which lattice fringes are observable.[27] Accordingly, of course, we cannot deny the existence of extended-chain crystals. Hence, in order to gain a better

understanding of the structure formation during crystallization under flow (oriented crystallization), the overall conformation of specific molecular chains in the material should be determined, probably by neutron scattering.

Non-uniformity of density generated in a polymer system which was already deformed or is now being deformed mechanically has been noted recently and termed stress-induced phase separation.[28] In this regard, an interesting result of simulation has been reported.[29] In this paper, a two-dimensional gel with a triangular network at its equilibrium state is assumed. A state with different strains which depend on the postion is given as an initial condition and the process of strain relaxation with time is simulated on the basis of dissipative molecular dynamics. As a result of such simulation, the existence of a metastable state is predicted, in which expanding and contracting phases coexist. If we regard the polymer melt as a gel in which the points of entanglement appear and disappear temporarily, we can understand the fact that in the relaxation process of a system deformed by elongational flow, there exists a metastable state which has spatial non-uniformity of density.

1.3.2 Oriented crystallization from solution

The previous subsection (1.3.1) introduced the fibrous precipitates grown from a solution of high density polyethylene, the so-called shish-kebab structure. It should be noted that the formation process of this structure was not a matter of common knowledge until comparatively recently. Here, the studies by McHugh *et al.*[30,31] on the formation process will be introduced. A 0.01wt% xylene solution of ultra-high molecular weight polyethylene ($M_W = 3 \times 10^6$) was flowed through a funnel-shaped pipe as illustrated in Fig. 1.5. A polarized light microscope was used to observe the formation process of the fibril at the tip of the seed crystal which was suspended in the pipe. The results indicated that a gel-like amorphous fibril, which had a higher concentration of the polymer and was non-oriented (or very weakly oriented), was formed first as a precursor and then crystallization synchronised with the emergence of birefringence which was induced by tensile stress exerted on the fibril through the fluid. From the time dependence of the change of the measured birefringence, the crystallization was expressed by the Avrami equation with an Avrami index of 2. McHugh *et al.*, therefore, concluded that the crystallization in question consisted of one-dimensional growth (growth in the flow direction) initiated by homogeneous nucleation. The formation of a precursor structure before crystallization bears a resemblance to structure formation in oriented crystallization from a melt and is therefore of great interest.

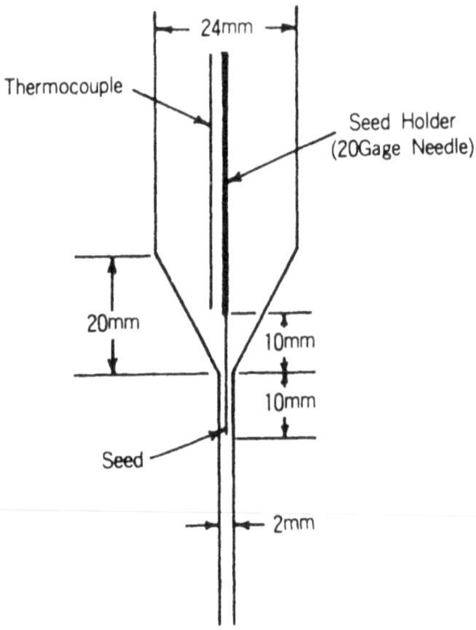

1.5 Apparatus for oriented crystallization of polyethylene from solution.[30]

1.3.3 Gelation and phase separation

Recently, theoretical and experimental studies on gelation and phase separation of polymer systems have flourished. According to experiments on a polyvinylalcohol (PVA)–water system by Komatsu *et al.*,[32] the sol–gel transition curve intersects the SD (spinodal decomposition) curve as shown in Fig. 1.6, and the phase diagram is partitioned into four areas:

(i) Area corresponding to a homogeneous sol state.
(ii) Area in which liquid–liquid separation takes place due to SD but gelation does not occur.
(iii) Area in which gelation takes place due to SD.
(iv) Area in which gelation takes place without any liquid–liquid separation.

The molecular orientation induced by the flow of such a solution, needless to say, shifts these curves. In dry spinning of the system in question, the path which the system follows on the phase diagram changes according to the spinning conditions and consequently the mode of structure formation and the structure itself in the spun filament will vary. As for structure formation in wet spinning of a three-component system, an excellent analysis has been presented in Reference 5.

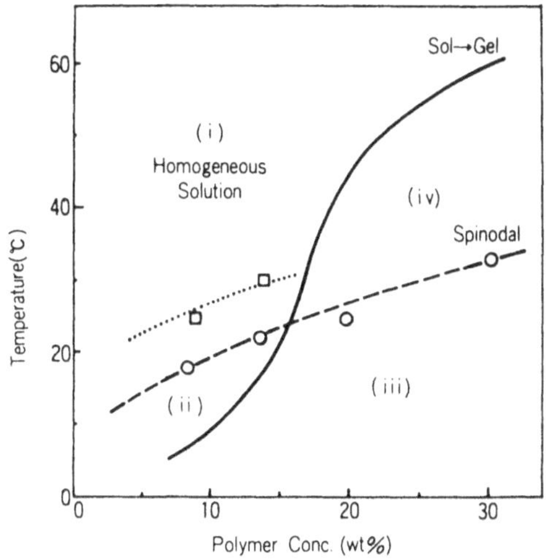

1.6 Phase diagram of PVA-water system for sol–gel transition (solid curve), spinodal (broken curve) and binodal (dotted curve).[32]

1.4 Fiber structure

For flexible polymers, uniaxial orientation is realised by the unfolding of their folded chains by stretching. Such orientation is also given to a rigid polymer by elongational flow of a liquid crystalline phase of the polymer; uniaxial orientation is a structural characteristic of fibers. The fiber structure in commercial fibers has been investigated by transmission electron microscopy (TEM), scanning electron microscopy (SEM), X-ray diffraction, and so on. Though only the outline fiber structure has been described so far, a more detailed picture of fiber structure obtained by recent TEM studies[33] will follow. In the context of the present chapter, the following section should only be regarded as a supplement. The authors hope, however, that it will assist readers in understanding the relationship between structure and physical properties of fibers.

1.4.1 Fiber structure of flexible chain polymers

1.4.1.1 Polyethylene (PE)

We first turn to a brief discussion of the fiber structure in an oriented thin film of PE.[33] The specimen was prepared by stretching, by a factor of 3–4, a thin film of PE spread on the surface of hot water and thereafter annealing it at 126°C. Figure 1.7(a) is a dark-field image using the 110 and 200 reflections on the equator (see the inset of Fig. 1.7), showing crystallites of about 20 nm length and width; crystallites which are properly oriented to give 110 or 200 reflections used for imaging appear as bright spots. These crystallites tend to align themselves in a line along the fiber axis (vertical direction), which suggests the presence of microfibrils in the specimen. Here and there, we often find domains where the crystallites surrounded by the amorphous part appear to have coherency of orientation over 200–300 nm along the fiber axis. Due to uniaxial orientation, the dark-field image (Fig. 1.7(b)) using the 002 reflection on the meridian is on the whole rather uniform in brightness and accordingly has lower contrast than Fig. 1.7(a).

Figure 1.7(c) is a phase-contrast image of the same specimen, which was taken at a fairly large amount of defocus. This figure clearly demonstrates wavy lamellae stacked in the direction of the fiber axis. Since the lateral dimension of these lamellae is much larger than that of the crystallites, it is deduced that each of the lamellae is a mosaic-like aggregate of crystallites. In conclusion, the oriented film of PE has two faces of fibrillar and lamellar structure. PE is too labile under electron irradiation to elucidate such seemingly contradictory features of structure by high-resolution TEM.

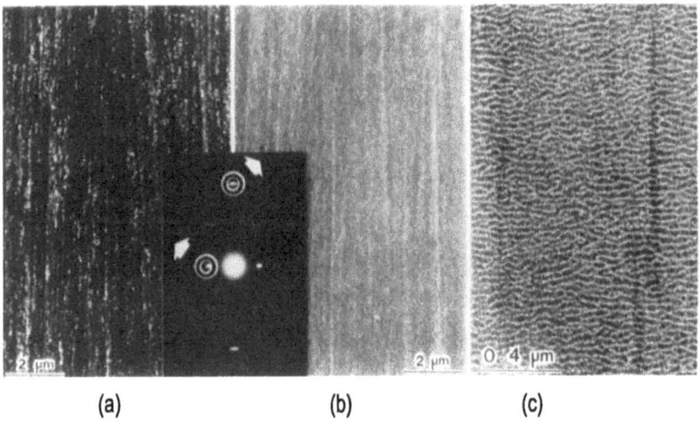

1.7 Dark-field image of polyethylene (PE). (a) Using the 110 and 200 reflections. (b) Using the 002 reflection. (c) Using the phase-contrast image. The inset is the corresponding electron diffraction pattern. The stretching direction is vertical.

1.4.1.2 Poly(aryl-ether-ether-ketone) (PEEK)

Next, we turn to the oriented thin film of PEEK.[34] PEEK is fairly resistant to electron bombardment and, thus, is a suitable material to take lattice images by high-resolution TEM. Several drops of a hot solution of PEEK in α-chloronaphthalene were sandwiched between two glass slides at 300°C. Just after evaporation of the solvent, a thin molten film of PEEK between the slides was oriented and crystallized by displacing one of the two slides. Figure 1.8(a) is the (110) lattice image of such a specimen, and Fig. 1.8(b) demonstrates the relationship between the lamellar structure and the domains in which (110) lattice fringes are recognized. This lamellar structure is illustrated by the phase-contrast image of the specimen, which was recorded at a large amount of underfocus just after an electron exposure for taking the (110) lattice image of Fig. 1.8(a). A fine crystallite connecting adjacent lamellae in the direction of the fiber axis is clearly seen; the authors have called this a 'tie-crystallite'. The authors are satisfied by the explanation that the correlation in orientation between crystallites which belong to a single microfibril is maintained by a tie-crystallite. Such tie-crystallites may pass through several lamellae. Tie-crystallites have also been recognised in the lattice images of uniaxially oriented thin films of PE and poly(4-methyl-1-pentene), whose images were taken at 4.2K using a cryogenic transmission electron microscope equipped with a superconducting objective lens.[35]

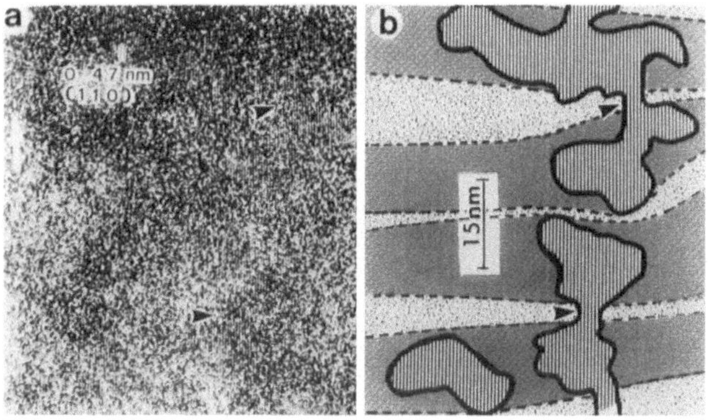

1.8 High resolution image of poly(aryl-ether-ether-ketone) (PEEK), showing tie-crystallites (arrowheads): (a) lattice image showing vertical shearing (see arrow); (b) schematic illustration showing the relationship between the lamellar structure and the domains in which (110) lattice fringes are observed in (a).

1.4.1.3 Isotactic polystyrene [i-PS]

Figure 1.9 is the high resolution image of an oriented thin film of *i*-PS,[27] which was prepared using a procedure developed by Tsuji *et al.* In the figure, (110) lattice fringes are running parallel to the direction of the fiber axis. The coherent region in which the fringes are observed is about 15 nm in width and attains to 200 nm in length. The fringes are slightly curved and partly faint, but the region seems to have no defects. This region corresponds to a *shish* in the shish-kebab structure which appears in oriented crystallization, and is considered to be a kind of tie-crystallite.

In Section 1.4.1.1, it was mentioned that the fiber structure of PE has two faces of lamellar and fibrillar structure. In ultra-drawn films/fibers of ultra-high molecular weight PE, one face, the lamellar structure, is lost

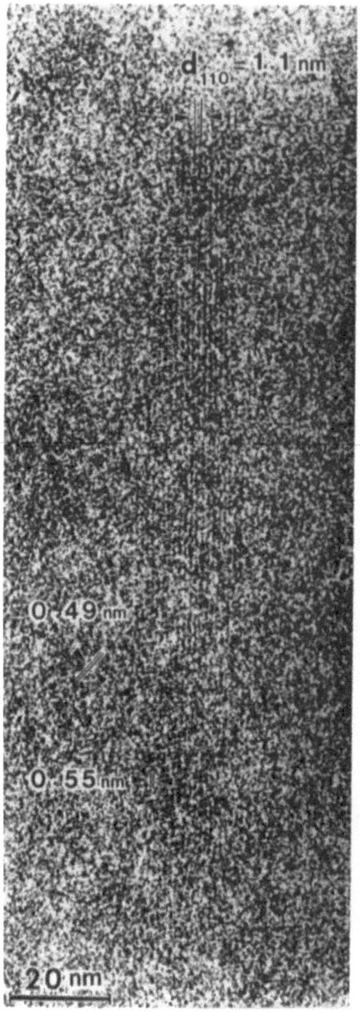

1.9 Structure observed in an oriented thin film of isotactic polystyrene (*i*-PS). The stretching direction is vertical.

and the other face, the fibrillar structure, having a great number of tie-crystallites, is enhanced. Accordingly, the films/fibers have excellent physical properties such as high modulus and high strength.

1.4.2 Fiber structure of rigid chain polymers

Fibers of rigid chain polymers are commonly produced by solidifying their liquid crystalline domains, in which extended chains are aligned parallel to each other, with all chains being oriented uniaxially owing to elongational flow working on the domains. As an example of fibers made in such a way, poly(*p*-phenylene terephthalamide) is discussed below.

1.4.2.1 Poly(p-phenylene terephthalamide) (PPTA)

A Kevlar fiber was annealed at 400°C under the condition of constant length, and then fibrillar fragments were obtained by tearing them off from the fiber. The fragments thus prepared were used as specimens for TEM. Figure 1.10(a) is the dark-field image of the specimen using the 006 meridional reflection, showing a periodic banded texture which consists of alternating bright and dark bands with a period of about 500 nm. Such a banded texture was also observed in longitudinal thin section[36,37] and in the fibrillar fragments[38] of PPTA. Figure 1.10(a) reveals that there exist microfibrils running through these bands along the direction of the fiber axis. The banded structure is produced by periodical bending of polymer chains. Figure 1.10(b) is the dark-field image of the same portion of the specimen that was used for Fig. 1.10(a), taken using the 110 and 200 reflections on the equator. The bright spots in Fig. 1.10(b), namely the crystallites oriented to give 110 or 200 reflections, are distributed seemingly in a random fashion. Careful inspection of the figure, however, shows that in some areas the crystallites are aligned along the microfibrils. The size of the crystallites is of the order of 10–20 nm both in width and in length, and is much smaller than that expected from

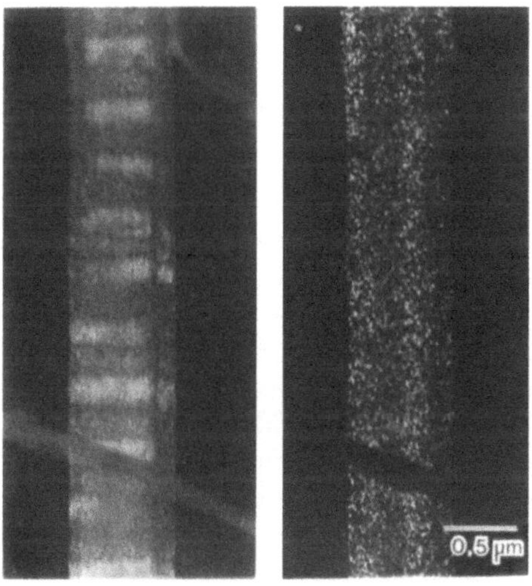

1.10 Dark-field images of a thin film of poly (*p*-phenylene terephthalamide) (PPTA): (a)–using the 006 reflection; (b)–using the 110 and 200 reflections. The fiber axis is in the vertical direction.

the nature of the rigid polymer chain and its chain length. This inconsistency seems to arise from twisting of microfibrils around their own chain axes. The dimension of the domains in which coherent (110) lattice fringes were observed by high-resolution TEM is also similar to the size of crystallites estimated above.

1.5 Simulation of high-speed spinning

In order to improve the productivity of fibers and to open up new avenues for their use, the process of high-speed melt spinning[9] has been proposed and developed as an attempt to produce highly oriented filaments by a single-stage process without any drawing after spinning. Since a commercial winder with a take-up speed of about 6000 m/min became available in the latter half of the 1970s, studies on high-speed spinning have been greatly advanced.

The features of high-speed spinning are as follows:

1 Polyethylene terephthalate (PET) does not crystallize appreciably in normal melt spinning, but does at a spinning speed of about 4000 m/min or more.
2 An abrupt necking-like change of diameter of the filaments during high-speed spinning is recognized and closely related with crystallization.[39,40]
3 The resultant spun filaments have a skin–core structure: the skin region is highly oriented and crystallized, but the core region has weak orientation and low crystallinity.
4 The spun filaments have rather high extensibility, and accordingly are utilized only for special purposes.
5 A spinning speed of 7000 m/min or more lowers the quality of spun filaments.

The object of this section is to demonstrate some results of a simulation of high-speed spinning which was carried out according to the procedure for the technological analysis of the spinning process as mentioned earlier in Section 1.2. First of all, the necking phenomenon will be discussed.

Figure 1.11 shows the change of diameter of the PET filament due to spinning speed in melt spinning experiments by Shimizu.[40] In these experiments, it was confirmed that the necking point moves up and down within certain limits, and that crystallization hardly occurs at all before necking, but increases very suddenly at the onset of necking. The real cause of this necking phenomenon is not yet known. One idea of the cause is based on the time dependence of elongational viscosity. Figure 1.12 is a plot of the elongational viscosity $\beta\ (t\ \dot{\varepsilon})$ in a non-steady state vs time in a situation where a melt of low density PE was elongated at a constant strain rate $\dot{\varepsilon}$.[41] $\beta\ (t\ \dot{\varepsilon})$ increases with time and attains a maximum. This indicates that the structural change due to elongational

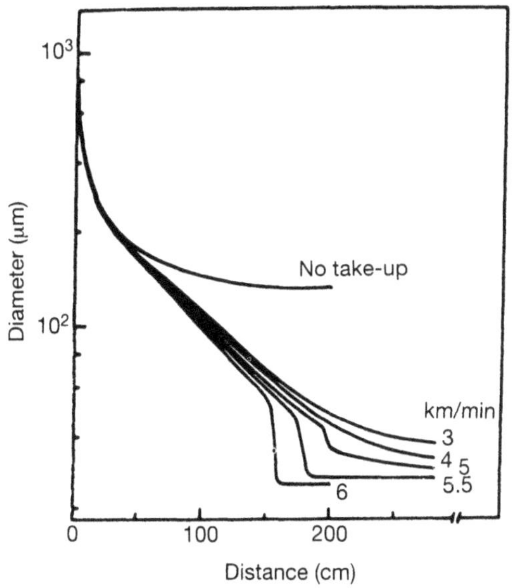

1.11 Change of diameter along the filament at various spinning speeds.[40]

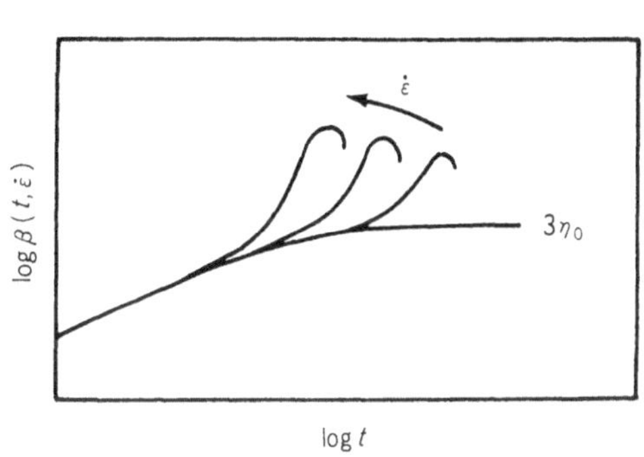

1.12 Change of viscosity in elongational flow.

flow has a complex influence on the viscosity, such that the viscosity depends on the strain itself as well as on the strain rate. Hence, it might be deduced that the viscosity reaches its maximum during the spinning process and this triggers the neck formation. On the other hand, another view that such a maximum might be due to experimental error and cannot exist has also been presented.[42] A second idea as to the cause of necking is the following. It was experimentally reported that when the ratio of elongational viscosity to shear viscosity exceeds 3, necking occurs.[43] According to this result, we can say that necking takes place when elongational strain attains a certain magnitude during spinning. A third idea comes from attention to the structural change which occurs before oriented crystallization. That is to say, the system itself is transformed into another state in which necking can readily take place (Shimizu named this state 'oriented mesophase').[40] For the moment, we cannot determine which of the above three ideas is true.

In order to start an actual simulation, numerical values or relations for all the related quantities are required. As a model polymer, PET with a molecular weight of about 23 000 was taken, and mathematical relations found in the literature were utilized. As for unknown numerical relations, appropriate assumptions were made, and a working hypothesis was also introduced for the mechanism of structural formation during spinning. Details of the calculation are to be found in Reference 10; only the results will be presented here.

Experimental results obtained for a mass output (W) of 2.55 g/min, extrusion temperature (spinning temperature) of 295°C and nozzle diameter of 0.3 mm are shown in Fig. 1.13, taken from Reference 16 and modified for convenience to be on the same scale as that of the results of simulation. In this simulation, the same spinning conditions as above were used, and a cooling condition was fixed at horizontal cooling air velocity of 40 cm/sec. Though the magnitude of initial tension must be given as an initial condition of calculation, this was roughly estimated by an established procedure in which crystallization is neglected. Since the take-up speed is to be determined in principle from the calculated result, it cannot be introduced as a spinning condition.

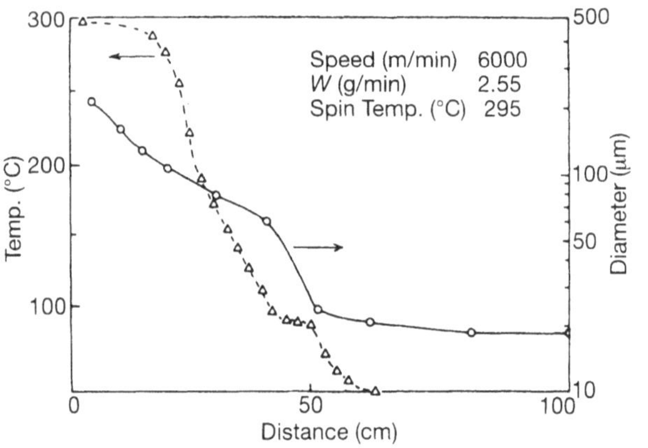

1.13 An example of a high-speed melt spinning experiment for polyethylene terephthalate.[16]

Figure 1.14 shows the change of diameter of the filament during spinning. In Fig. 1.15, changes in diameter, surface temperature and crystallinity along the spinning path are shown in the same figure for the case with a spinning speed of 5700 m/min. In Fig. 1.16, radial distributions of relative crystallinity at several distances from the spinneret are illustrated for a spinning speed of 5700 m/min. This figure

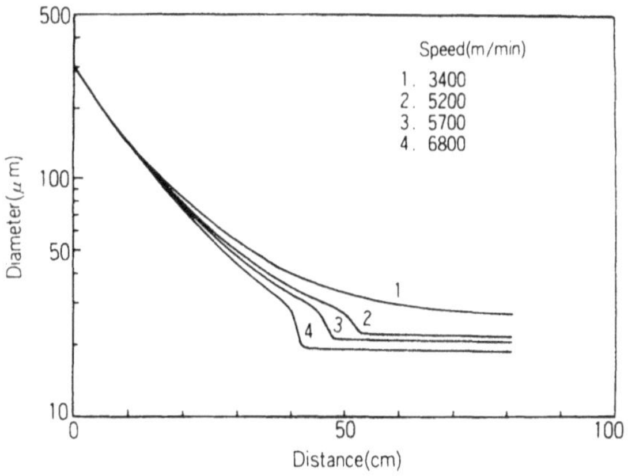

1.14 Thinning of the filament (simulation).

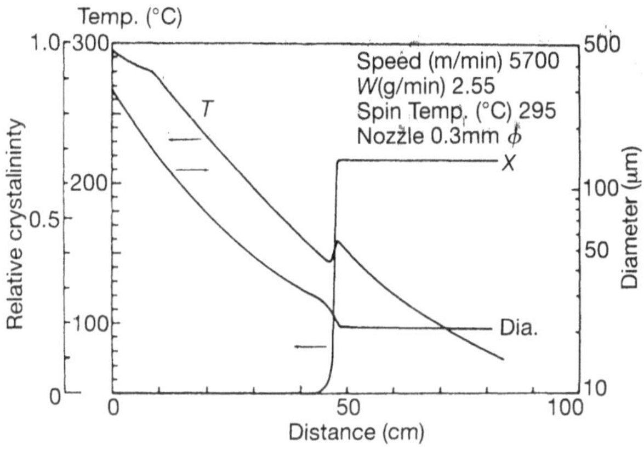

1.15 Changes in diameter, surface temperature and crystallinity along the spinning path (simulation).

clearly reveals the behavior of crystallization advancing from the filament surface to its inside. The lower curve in Fig. 1.17 is the plot of the calculated tensile strength at the filament surface against temperature. The upper curve in the same figure shows the experimental necking stress (necking tension per cross-sectional area just before the onset of necking) for an unoriented PET film annealed for 3 hrs at 120°C. It is deduced that the real necking stress during spinning should be rather smaller than the experimental necking stress mentioned above. However, the fact that the two curves in Fig. 1.17 intersect each other at a point corresponding to about 150°C definitely predicts the appearance of necking around this point during high-speed melt spinning.

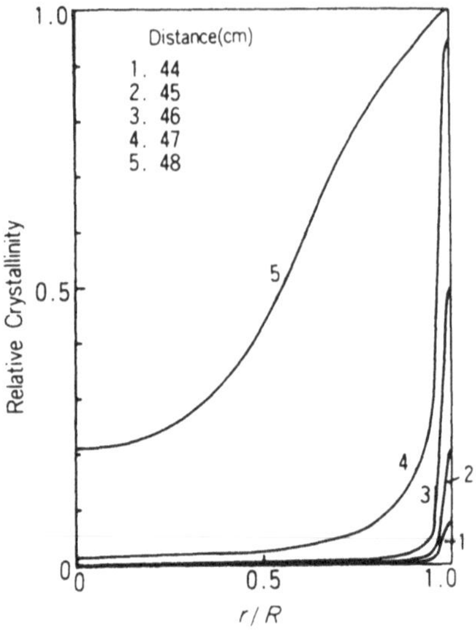

1.16 Radial distributions of relative crystallinity at five distances from the spinneret at a spinning speed of 5700 m/min (simulation).

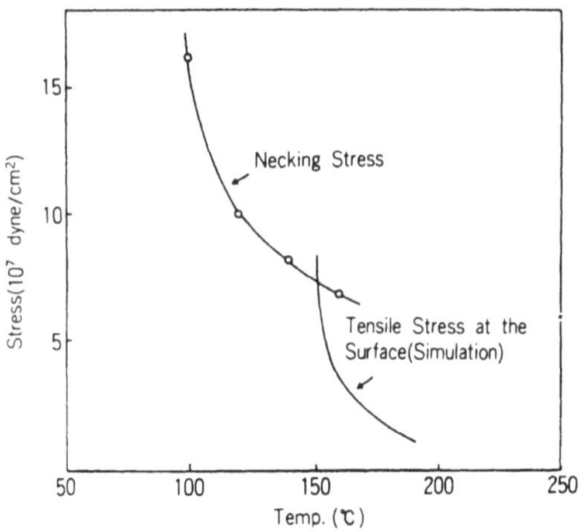

1.17 Necking stress of unoriented and annealed film of polyethylene terephthalate (PET) (observed) and the tensile stress at the filament surface (simulation, spinning speed = 5700 m/min).

1.6 Concluding remarks

In this chapter, a mathematical formulation for technological analysis of the spinning process and the features of structure formation in oriented crystallization were described. Their application to the simulation of high-speed melt spinning was also shown. The purpose of such simulation is not to obtain a quantitative agreement between calculated values of physical properties and those observed in the real spinning process, but to determine what information is lacking for an exact simulation and to build a framework for calculations. Lack of understanding of oriented crystallization is a serious deficiency of the present simulation. When only primary crystallization, which can be expressed with an Avrami equation is taken into consideration, it can be predicted that crystallization will take place very suddenly and the filament temperature will rise quickly by 10° or more. Practically, however, it is probable that the crystallization rate is not so great in the latter half of the crystallization process where secondary crystallization plays an important role. This subject should be studied in the future, as should the temperature dependence of crystallization rate and its dependence on the degree of orientation. For a full understanding of the necking phenomenon, precise on-line measurements and further structural investigations are needed.

References

1 Kase S and Matsuo T, *J. Polym. Sci.*, **A3**, 2541, 1965.
2 Kase S and Matsuo T, *J. Appl. Polym. Sci.*, **11**, 251, 1967.
3 Katayama K, Amano T and Nakamura K , *Kolloid-Z.Z. Polym.*, **226**, 125, 1968.
4 *Formation of Fibers and Development of their Structure: (I) Melt Spinning*, Ed. The Society of Fiber Science and Technology, Japan, Kagaku-dojin, Kyoto, 1969.
5 *Formation of Fibers and Development of their Structure: (II) Wet Spinning and Dry Spinning*, Ed. The Society of Fiber Science and Technology, Japan, Kagaku-dojin, Kyoto, 1970.
6 Ziabicki A, 'Physical fundamentals of the fiber spinning process, in *Man Made Fibers*, Vol. 1, Ed. H F Mark, S M Atlas and E Cernia, Interscience, New York, 1967.
7 Ziabicki A, *Fundamentals of Fibre Formation*, Wiley, London, 1976.
8 White J L, *Polym. Rev.*, **1**, 297, 1981.
9 Ziabicki A and Kawai H eds., *High Speed Fiber Spinning: Science and Engineering Aspects*, Wiley, New York, 1985.
10 Katayama K and Yoon M-G, *Sen-i Gakkaishi*, **38**, P-434, 1982.
11 Kawai T, Iguchi M, and Tonami H, *Kolloid-Z. Z. Polym.*, **221**, 28, 1967.
12 McHugh A J and Yung W S, *J. Polym. Sci.Phys.*, **27**, 431, 1989.
13 Smith F S and Steward R D, *Polymer*, **15**, 283, 1974.
14 Nakamura K, Watanage T, Katayama K and Amano T, *J. Appl. Polym. Sci.*, **16**, 1077, 1972.
15 Nakamura K, Katayama K and Amano T, *J. Appl. Polym. Sci.*, **17**, 1031, 1973.
16 Matsui M, *Sen-i Gakkaishi*, **38**, P-508, 1982.

17 Treloar L R G, *The Principles of Rubber Elasticity*, 3rd Edn, Clarendon Press, Oxford, 1975.
18 Ziabicki A and Jarecki L, *Colloid & Polym. Sci.*, **256**, 332, 1978.
19 Ohzawa Y, Nagano Y and Matsuo T, *J. Appl. Polym. Sci.*, **13**, 257, 1969.
20 Ohzawa Y and Nagano Y, *J. Appl. Polym. Sci.*, **14**, 1879, 1970.
21 Pennings A J and Kiel A M, *Kolloid-Z.Z. Polym.*, **205**, 160, 1965.
22 Pennings A J, van der Mark J M A A and Kiel A M, *Kolloid-Z.Z. Polym.*, **237**, 336, 1970.
23 Katayama K, *J. Soc. Rheol., Jpn.*, **4**, 96, 1976.
24 Keller A and Machin M J, *J. Macromol. Sci.Phys.*, **B1**, 41, 1967.
25 Hill M J and Keller A, *J. Macromol, Sci.Phys.*, **B5**, 591, 1971.
26 Amano T, Kajita S and Katayama K, *Progr. Colloid & Polym. Sci.*, **58**, 108, 1975.
27 Tsuji M, Uemura A, Ohara M, Kawaguchi A, Katayama K and Petermann J, *Sen-i Gakkaishi*, **42**, T-580, 1986.
28 Rangel-Nafaile C, Metzner A B and Wissbrun K F, *Macromolecules*, **17**, 1187, 1984.
29 Sekimoto K, Suematsu N and Kawasaki K, *Phys. Rev. A*, **39**, 4912, 1989.
30 Rietveld J. and McHugh A J, *J. Polym. Sci.Phys.*, **23**, 2339, 1985.
31 McHugh A J and Spevacek J A, *J. Polym. Sci. C*, **25**, 105, 1987.
32 Komatsu M, Inoue T and Miyasaka K, *J. Polym. Sci. B*, **24**, 303, 1986.
33 Katayama K, Isoda S, Tsuji M, Ohara M and Kawaguchi A, *Bull. Inst. Chem. Res., Kyoto Univ.*, **62**, 198, 1984.
34 Kawamura H, Tsuji M, Kawaguchi A and Katayama K, *Bull. Inst. Chem. Res., Kyoto Univ.*, **68**, 41, 1990.
35 Tsuji M, Tosaka M, Kawaguchi A, Katayama K and Iwatsuki M, *Sen-i Gakkaishi*, **48**, 384, 1992.
36 Dobb M G, Johnson D J and Saville B P, *J. Polym. Sci., Polym. Symp.*, **58**, 237, 1977.
37 Dobb M G, Johnson D J and Saville B P, *J. Polym. Sci. Polym. Phys. Ed.*, **15**, 2201, 1977.
38 Takahashi T, Miura M, and Sakurai K, *J. Appl. Polym. Sci.*, **28**, 579, 1983.
39 Perez G and Lecluse C, *Proceedings of the 18th International Man-Made Fibre Conference*, Dornbirn, Lenzing AG, 1979.
40 Shimizu J, *Sen-i Gakkaishi*, **38**, P-499, 1982.
41 Raible T and Meissner J, 'Uniaxial extensional experiments with large strains performed with low density polyethylene (LDPE)', in *Rheology, Vol. 2: Fluids* ed. G Astarita, G Marrucci and L Nicolais, Plenum Press, New York, 425, 1980.
42 Masuda T, private communication.
43 Jin Xia, Okui N, Umemoto S and Sakai T, *Polym. Prepr., Jpn.*, **39**, 3683, 1990.

2

Melt spinning

Yasuhiro Murase
Teijin Ltd, Ibaraki, Japan
Akihiko Nagai
Seitoku University, Tokyo, Japan

2.1 Introduction

This chapter describes the ultra-high speed spinning of polyethylene terephthalate (PET) as a typical example of melt-spun synthetic yarn. PET yarn has been produced by a conventional spinning and drawing process since 1958 when its domestic production in Japan started. In that process, the PET melt is extruded and then wound up at a speed of 1200 m/min. The resulting undrawn yarn (UDY) is then drawn by 3–5 times and heat-treated to give the fully oriented yarn (FOY or DY) as shown schematically in Fig. 2.1. PET drawn yarn is mostly produced according to this system.

The drawing process can be omitted in high-speed spinning, and the drawn yarn be made economically in one step. This idea was patented by Du Pont[1] in the 1950s. However, the full development of such a one-step technology required development of the high-speed winder and this did

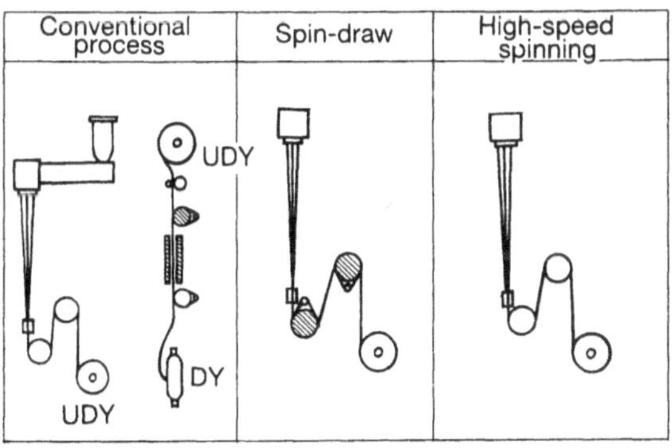

2.1 Schematic production system for PET yarn.

not occur until 1988 when commercial high-speed spinning at over 6000 m/min started.

The direct spin draw process was developed conventionally in the 1960s by coupling the spinning and drawing processes in series (Fig. 2.1). As the demand for crimped yarn increased in the 1970s, the drawing and texturing processes were combined into one process, and a new spinning process was developed to produce partially oriented yarn (POY) with a spinning speed of 3000–3500 m/min. to produce the feed stock for integrated draw-texturing.

The spinning speed increased as winder performance improved in the early 1970s, and this development encouraged the investigation of high-speed spinning. Ueda and Kanatsuna, for example, reported the fiber structure of nylon 6 made by high-speed spinning at up to 9800 m/min in 1971.[2] For ten years from 1975, the high-speed spinning process was extensively investigated by various researchers including Shimizu *et al.* who reported a series of high-speed spinning results on polyester, nylon and polyolefin.[3] In 1983, the Association for Efficient Synthetic Fibre Technology was established in Japan with a scheme of conditional loans for research and development of innovative technologies under the Ministry of International Trade and Industry; its spinning section coordinated a project for high-speed spinning of polyester at 9000–14 000 m/min. These investigations have revealed the optimum conditions for high-speed spinning and the mechanism of the fiber structure formation during the process. Today, high-speed spinning at 6000–8000 m/min is in commercial operation for the production of synthetic yarns such as nylon and polyester.

This chapter outlines the structure and physical properties of polyester filament yarn obtained by such high-speed spinning, the mechanism of the fiber structure formation during the process, and the commercial applications.

2.2 Mechanical properties and structure of high-speed spun yarn

This section summarizes the structure and physical properties of PET yarns produced by high-speed spinning.

2.2.1 Yarn quality in general

Figures 2.2 (a) and (b) show the spinning speed dependence of the tenacity and elongation of polyester yarn, respectively. The stress–strain curve is shown in Fig. 2.3 for yarns spun at various velocities. The yarn tenacity increases and its elongation at break decreases as the spinning speed increases. When the spinning speed exceeds 5000 m/min, the elongation falls below 70%, and the yield point in the stress–strain curve becomes less distinctive, as in conventional FOY. The tenacity exhibits a

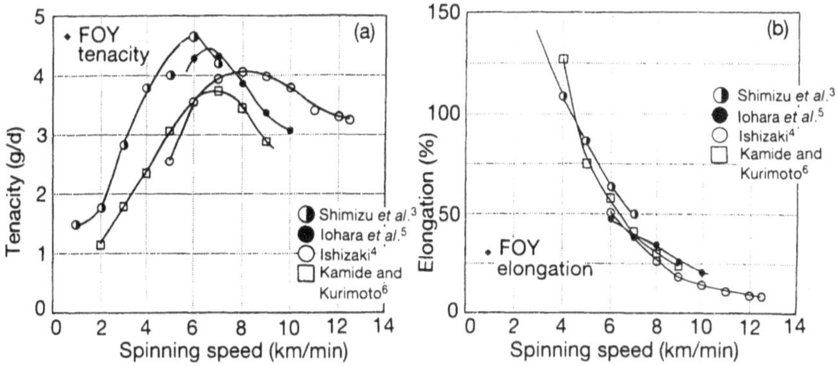

2.2 (a) Dependence of tenacity on spinning speed, and (b) dependence of elongation on spinning speed.[3–6]

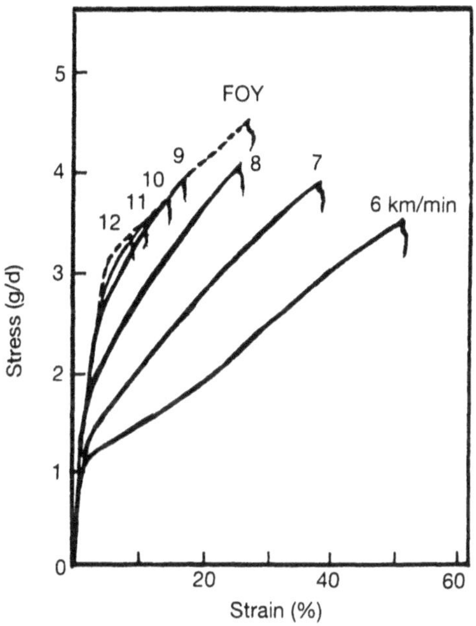

2.3 Stress–strain curve at various spinning speeds.

maximum at a spinning speed of 6000–7000 m/min and then decreases as seen in Fig. 2.2 (a). (Data from Ishizaki *et al.*[4] exhibit a maximum at a higher spinning speed of 8000 m/min, also shown in Fig. 2.2 (a). This may be due to a difference in the PET chemical structure, since a modified PET was used in order to improve the spinnability.) The tenacity of the high-speed-spun yarn is in the range 3.8–4.7 g/d, which is slightly lower than that of FOY. The elongation decreases with spinning speed, and becomes less than 25% at a spinning speed of over 8000 m/min. Young's modulus increases abruptly at a spinning speed of 3000–4000 m/min (see Fig. 2.4), suggesting a large change of fiber structure at this point. Although Young's modulus increases continually up to a spinning speed of over 6000 m/min, according to the results of Shimizu *et al.*[3] and Kamide *et al.*,[6] it may decrease when the spinning speed is further increased, since it tends to fall at spinning speeds of over 8000 m/min, like the tenacity, according to the results of Fujimoto.[7] The Young's modulus of high-speed-spun PET yarn reaches 100 g/d, as high as that of FOY.

Figure 2.5 shows the spinning speed dependence of the thermal shrinkage in boiling water. The thermal shrinkage exhibits a maximum (around 60%) at a spinning speed of 2000–3000 m/min, and then decreases to as low as 2–3% at a spinning speed of over 6000 m/min. Thus the thermal shrinkage stability is good in high-speed-spun PET yarns.

The dye pickup for 60-minute dyeing was examined with two dyes of different molecular weights as a function of the spinning speed (Fig. 2.6). Both dyes exhibited the minimum pickup at a spinning speed of 5000 m/min, with an increase at higher spinning speed. The dye pickup of the PET yarn spun at 9000 m/min was found to be around 70% when it was dyed with a disperse dye (Resolin Blue FBL) under atmospheric pressure at 100°C. This value is not as high as the 85% found for FOY dyed with the same liquor ratio under high pressure at 130°C, but the dyeability of the high-speed-spun PET yarn is good, considering that the value of 70% was obtained at only 100°C.

The characteristics of high-speed-spun PET yarn can be summarized as:

1 The tenacity and Young's modulus are slightly lower than those of FOY.

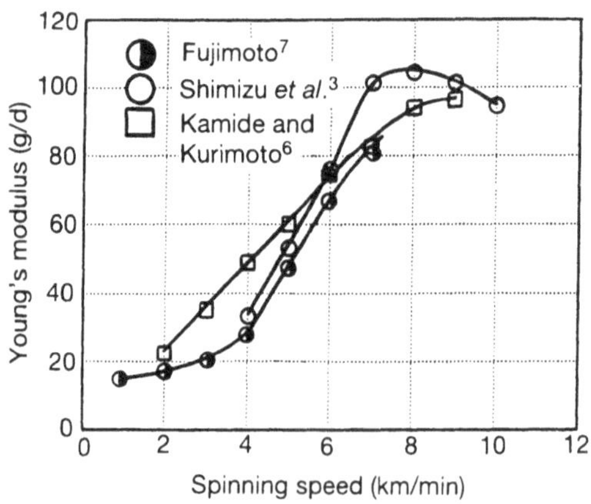

2.4 Dependence of Young's modulus on spinning speed.[3,6,7]

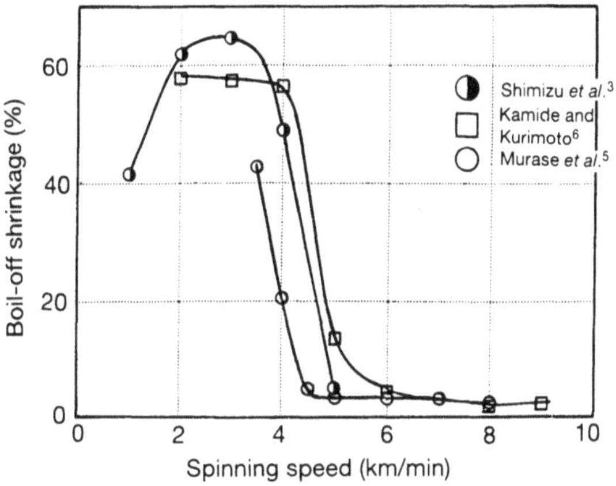

2.5 Dependence of boil-off shrinkage on spinning speed.[3,5,6]

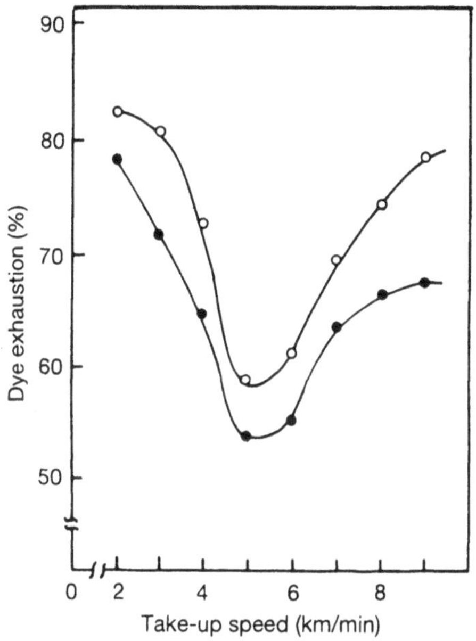

2.6 Dependence of dye exhausion on spinning speed[6] (dyed for 60 min with Diacelliton Fast Orange GL ○ and Resolin Blue TBL ●)

2 The elongation of the high-speed-spun yarn exhibiting the maximum tenacity is higher than that of FOY.
3 The thermal shrinkage is low, and the thermal shrinkage stability is good.
4 The dyeability is good.

2.2.2 Spinning speed dependence of physical properties of high-speed-spun yarn

Polymer chains orient and crystallize, and the fiber structure develops as the spinning speed increases.

The birefringence Δn and the degree of crystallinity (calculated from the density) depend on the spinning speed as shown in Fig. 2.7 and 2.8, respectively. Δn increases slowly at the beginning (up to a spinning speed of 2000 m/min) and then sharply above a spinning speed of 3000 m/min. It reaches a maximum value of $\Delta n = 0.12$ at a spinning speed of 7000 m/min, but decreases to $\Delta n = 0.08$ at a spinning speed of 9000 m/min. This is the same value as at a spinning speed of 5000 m/min. Similarly, the degree of crystallinity reaches a maximum value of 45–50% at a spinning speed of 7000 m/min (Fig. 2.8), and falls to below 40% at a spinning speed of 10 000 m/min. Although Δn starts increasing at a spinning speed of 2000 m/min,[4,8] the degree of crystallinity hardly changes up to the spinning speed of 4000 m/min and then increases suddenly.[4,8,9] Shimizu *et al.*[3] found a break point at $\Delta n = 0.055$ in the density–Δn relation as shown in Fig. 2.9. These results show that the molecular orientation improves with spinning speed, with little change in crystallinity, at spinning speeds below 4000 m/min, and orientation-induced crystallization takes place at spinning speeds over 4000 m/min. Δn and the

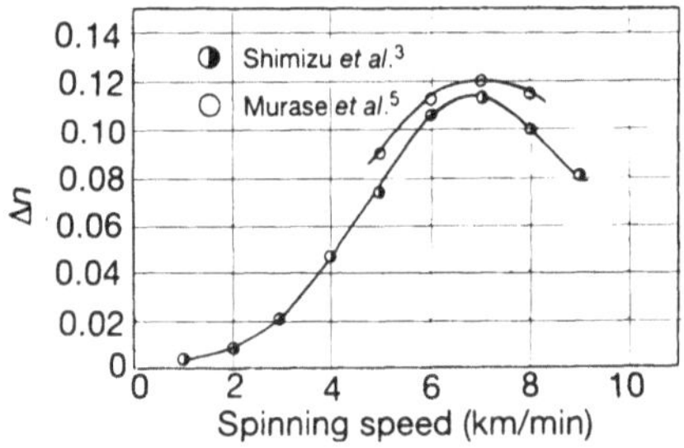

2.7 Dependence of birefringence Δn on spinning speed.[4,8,9]

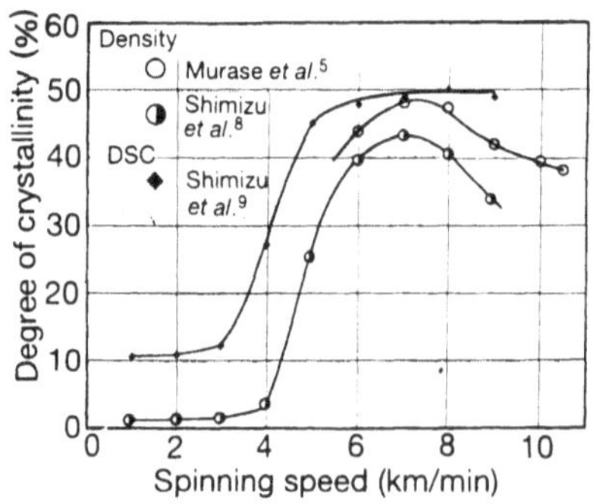

2.8 Dependence of degree of crystallinity on spinning speed.[4,8,9]

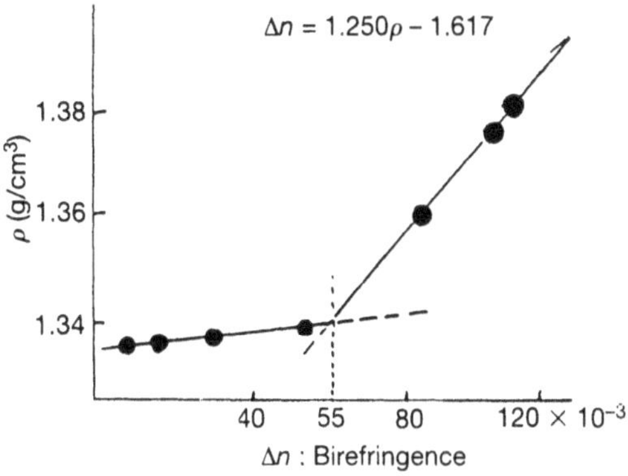

2.9 Relation between density and birefringence.[8,9]

degree of crystallinity (evaluated by density) exhibit maxima at 7000 m/min resulting from an inhomogeneous structure in the fiber cross-section due to the distribution of molecular orientation in the radial direction. The degree of crystallinity evaluated by DSC, shown in Fig. 2.8, exhibits a similar spinning speed dependence to that exhibited by density up to 7000 m/min, but no decrease above 7000 m/min. This difference is due to the presence of voids in the fiber surface and the skin–core structure of the fiber. The DSC crystallinity also grows at spinning speeds over 4000 m/min.

The chain orientation can be evaluated separately in the crystalline and amorphous phases from the birefringence. The total birefringence is given in terms of the birefringence in the crystalline phase Δn_c, the birefringence in the amorphous phase Δn_a and the volume fraction degree of crystallization x as

$$\Delta n = \Delta n_c x + \Delta n_a (1 - x) \qquad [2.1]$$

$$\Delta n_c = fc \Delta n_{c_0} \qquad [2.2]$$

where fc denotes the degree of orientation in the crystalline phase evaluated by the X-ray diffraction method and the limiting birefringence in the crystalline phase is calculated as $\Delta n_{c_0} = 0.212$. Figure 2.10 shows the spinning speed dependence of Δn_c and Δn_a evaluated from equation 2.1. Although the birefringence in the crystalline phase Δn_c cannot be evaluated unless the crystalline phase is formed (i.e., at spinning speeds over 4000 m/min), Δn_c is already as high as 0.16 at a spinning speed of 4000 m/min and increases to over 0.20 at a spinning speed of 8000 m/min, suggesting that the chain orientation is closely related to crystallization. The birefringence in the amorphous phase Δn_a reaches a maximum value of 0.05 at a spinning speed of 6000 m/min, and then gradually decreases with increasing spinning speed. The total birefringence Δn of the fiber

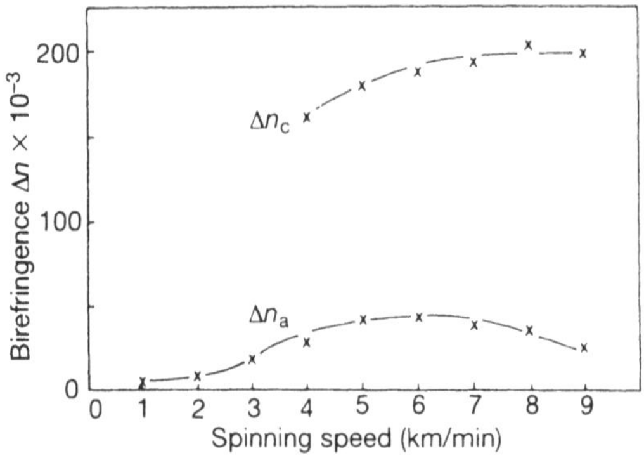

2.10 Crystalline orientation Δn_c and amorphous orientation Δn_a as a function of spinning speed.[8,9]

reaches a maximum value of 0.12 at a spinning speed of 7000 m/min, (Fig. 2.7). This is lower than the value of 0.15 obtained for the conventional FOY (fully oriented yarn), probably due to relatively low orientation in the amorphous phase. To sum up, the high-speed-spun yarn is characterized by high crystalline orientation and low amorphous orientation.

2.2.3 Skin–core structure of fiber

Figure 2.11 shows the change in birefringence Δn of a filament radially from centre to surface. Δn exhibits only a small variation radially up to a spinning speed of 8000 m/min, but decreases suddenly at the centre of the fiber when the spinning speed exceeds 9000 m/min. This means that the ultra-high-speed-spun yarn (spinning speed over 8000 m/min) has a

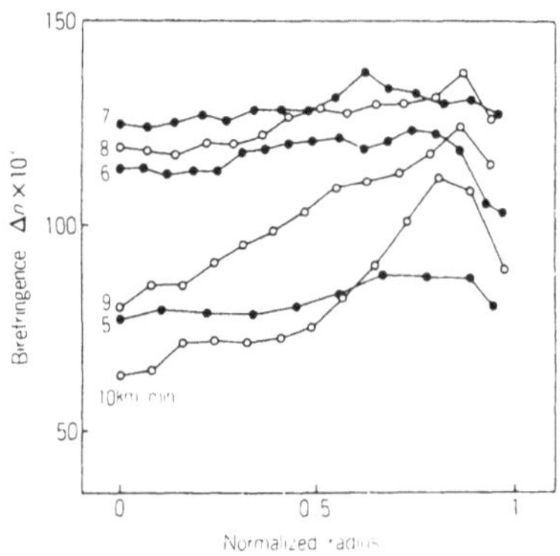

2.11 Distribution of Δn in cross-section of PET filament spun at various spinning speeds.[7]

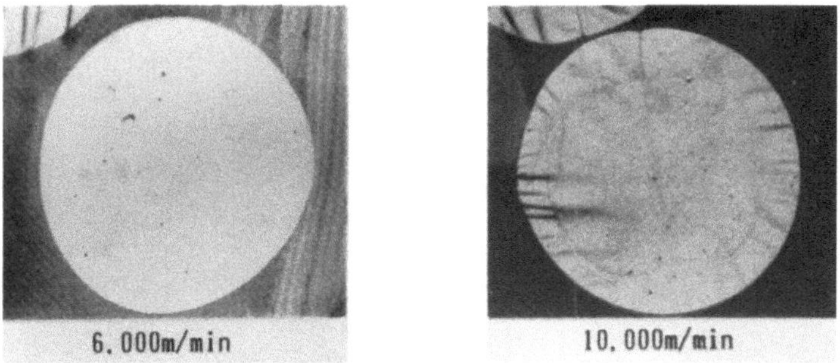

2.12 Transmission electron microscopic view of high-speed-spun filament cross-section.[7,10]

radially inhomogeneous structure with a high Δn at the surface and a low Δn at the centre. In Fig. 2.12, where filaments spun at 6000 m/min and 10 000 m/min were cross-sectioned and dyed by osmic acid, the transmission electron microscope (TEM) reveals a clear boundary between skin and core in the filament cross-section at the higher speed. Δn was found to be about 0.06 in the core of this filament. This value is close to the value observed for POY, so the core is considered to assume a POY-like structure.

2.2.4 Spinning speed dependence of structure in the crystalline phase

The Bragg spacings of high-speed-spun filaments are summarized in Table 2.1. The (010) spacing and (100) spacing decrease with increasing spinning speed, but the ($\bar{1}05$) exhibits no variation. This suggests that the crystal lattice of the ultra-high-speed-spun filament is compact in the radial direction with respect to the fiber axis. The crystal density evaluated from the lattice constants increases with spinning speeds up to 7000 m/min, and the crystal packing improves accordingly (Fig. 2.13). However, the radial structure becomes inhomogeneous and the crystal density deteriorates at higher spinning speeds of 8000 m/min and 9000 m/min.

Table 2.1 Spinning speed dependence of the Bragg spacings

	Bragg spacing				
	(010)	(100)	(200)	($\bar{1}03$)	($\bar{1}05$)
5000m/min	5.07 Å	3.48 Å	(3.47) Å	3.40 Å	2.11Å
6000m/min	5.01 Å	3.44 Å	(3.44) Å	3.37 Å	2.10Å
7000m/min	4.98 Å	3.42 Å	(3.42) Å	3.37 Å	2.10Å
8000m/min	4.99 Å	3.42 Å	(3.42) Å	3.38 Å	2.10Å
9000m/min	4.99 Å	3.42 Å	(3.42) Å	3.38 Å	2.10Å
Drawn and heat-treated	5.04 Å	3.46 Å	(3.45) Å	3.36 Å	2.10Å

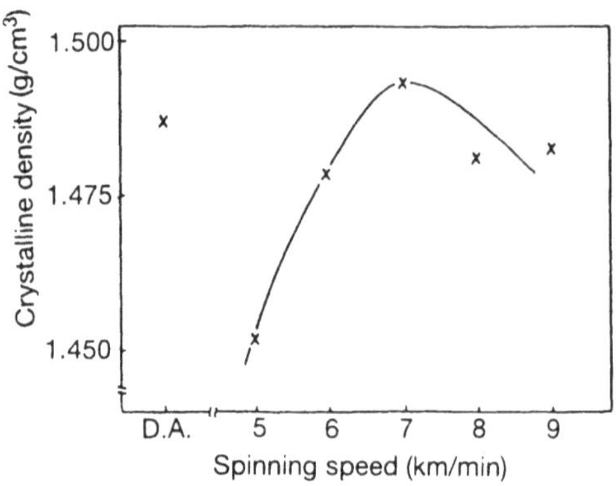

2.13 Dependence of crystalline density on spinning speed.

Table 2.2 Spinning speed dependence of crystal size

	According to Scherrer			According to Hosemann			
	(010)	(100)	(200)	(100),(200)	g_{II}	($\bar{1}$03),($\bar{1}$05)	g_{II}
5000m/min	49.4 Å [9.7]	29.7 Å [7.5]	29.5 Å [8.5]	(29.5) Å	–	66.8 Å [6.2]	1.12%
6000m/min	59.6 Å [11.9]	49.5 Å [12.6]	46.0 Å [13.4]	47.8 Å [13.9]	4.5%	91.6 Å [8.5]	0.75%
7000m/min	67.2 Å [13.5]	53.9 Å [13.8]	53.3 Å [15.6]	55.8 Å [16.3]	4.4%	102.9 Å [9.6]	0.71%
8000m/min	72.1 Å [14.4]	60.9 Å [15.4]	63.9 Å [18.7]	69.0 Å [20.2]	4.5%	115.6 Å [10.8]	0.61%
9000m/min	81.1 Å [16.3]	69.3 Å [17.6]	65.2 Å [19.1]	72.5 Å [21.2]	4.8%	90.0 Å [8.4]	0.53%
Drawn and heat-treated	64.1 Å [12.7]	47.8 Å [12.3]	41.0 Å [11.9]	42.1 Å [12.3]	4.4%	60.1 Å [5.6]	0.40%

The crystallite size also increases with spinning speed. Table 2.2 lists the crystallite dimensions estimated by Shimizu *et al.*[8,11] according to the equations of Scherrer and Hosemann. The crystal size in the *c*-axis direction, evaluated from the ($\bar{1}$03) and ($\bar{1}$05) planes, is already larger at a spinning speed of 5000 m/min than that of conventional drawn and heat-treated filaments, and increases further with spinning speeds up to 8000 m/min, where the crystal size is twice as large as that of the drawn and heat-treated filament. The crystal size becomes smaller at a spinning speed of 9000 m/min, but is still larger than for the drawn and heat-treated filament. The crystal size in the radial direction of the fiber axis was estimated as the dimension perpendicular to each of the (100), (010) and (100) planes, and was found to increase monotonically with the spinning speed. Here the crystal size is smaller up to a spinning speed of 6000 m/min than that of the conventional drawn and heat-treated filament. The values in square brackets in Table 2.2 are calculated by dividing crystal sizes by corresponding unit cell spacings, and denote the number of unit cells stacked in respective directions. The crystal is composed of approximately 15 unit cells at a spinning speed of 7000 m/min. The number of unit cells in the [100] direction, i.e., in the direction perpendicular to the (100) plane, is larger than that in other directions, and amounts to 20 at a spinning speed above 8000 m/min. Considering that the aromatic ring lies on the (100) plane of the unit cell, the crystallization proceeds in the particular direction in which the aromatic rings are stacked.

The DSC curve of polyester filaments is composed of exothermic (due to crystallization at a lower temperature) and endothermic (due to melting at a higher temperature) peaks as shown in Fig. 2.14. The crystallization peak shifts to a lower temperature at a spinning speed of 6000 m/min, and the peak is weak. When orientation-induced crystallization takes place at a spinning speed of 8000 m/min, the crystallization peak disappears. The peak temperature due to melting shifts to a higher temperature with the spinning speed, and a shoulder appears at a lower temperature than the main melting peak when the spinning speed exceeds 10 000 m/min. This shoulder is considered to be due to the melting of the core part in the skin–core structure of the filament. An exothermic peak due to crystallization is also observed at about 105°C in conjunction with the appearance of the shoulder. Kikutani *et al.* speculate that this exothermic peak results from a POY-like structure formed in the core part at spinning speeds higher than 10 000 m/min. Figure 2.15 is based on

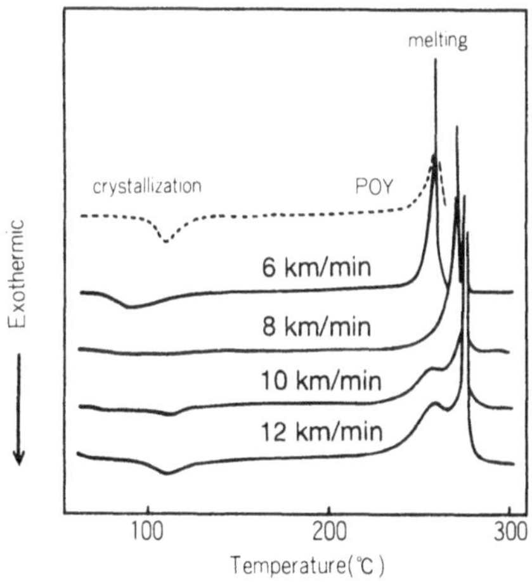

2.14 DSC thermograms of PET filaments spun at various spinning speeds.[5]

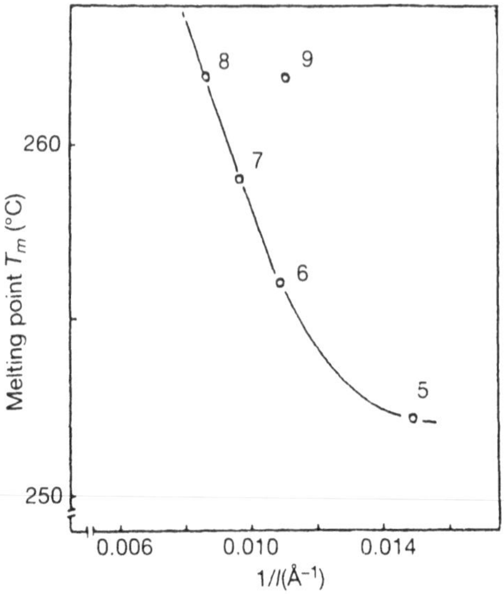

2.15 Melting peak temperature (T_m) as a function of crystal thickness.[8,11]

the results of Shimizu *et al.* where the melting peak temperature is plotted against the inverse of the crystal thickness estimated by wide-angle X-ray scattering. The melting point increases with the crystal thickness, and is extrapolated to 285°C which corresponds to the equilibrium melting temperature.

2.2.5 High-order structure of high-speed-spun filament

Figure 2.16 shows schematically the small-angle X-ray scattering from high-speed-spun PET filaments. An X-shaped four-point image observed at a spinning speed of 5000–6000 m/min starts deforming when the spinning speed exceeds 7000 m/min. The scattering image consists of broad two-point images expanding towards the meridional direction in an inverse triangular shape and a streak in the equatorial direction at a spinning speed of 8000–9000 m/min. Shimizu *et al.* calculated the scattered intensity distribution from a model composed of crystallites arranged in a specific manner according to the theory of Tsvankin, and simulated the expected small-angle X-ray scattering profile. The results are shown in Fig. 2.17. Here a single block is composed of six small crystallites arranged in a chequered pattern and the interspatial distance between blocks is large at a spinning speed of 5000 m/min; this is compared with a single block composed of two larger crystallites with a shorter interspatial distance at a spinning speed of 9000 m/min. The amorphous region binds two crystallite blocks in the horizontal direction, and forms a microfibril structure.

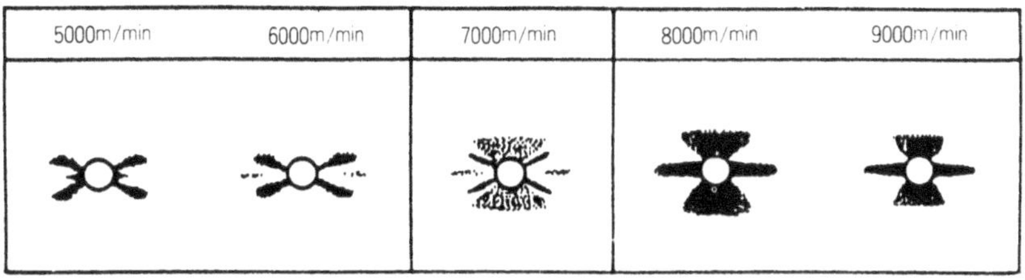

2.16 Small-angle X-ray diffraction pattern (schematic).

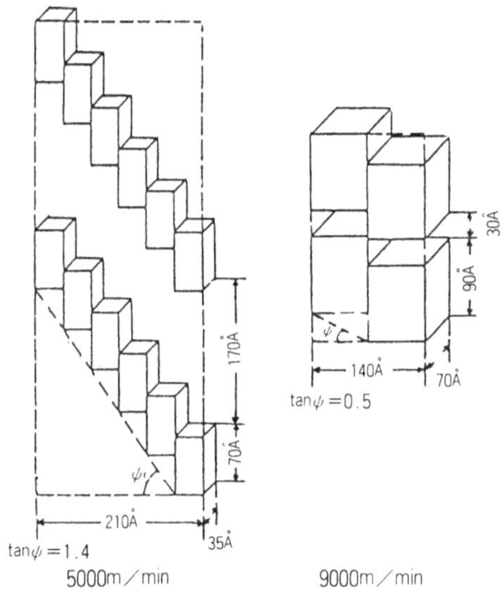

2.17 Fine structure model of filaments spun at 5000 m/min and 9000 m/min (from Shimizu *et al.*[8,11]).

The crystalline fraction (i.e. the crystallite thickness/the long period) evaluated from the two-phase model of repeating crystalline and amorphous phases is about 28% (cf. x = 26% estimated from the density) for the filament spun at 5000 m/min, but this value becomes too high (75%) at 9000 m/min (cf. x = 34% estimated from the density). Shimizu *et al.* concluded that this discrepancy was caused by inter-microfibrillar voids which were not taken into account in the simulation.

Ishizaki *et al.*[5,7] proposed a fine structure model based on the results of the small-angle X-ray scattering profile along the equatorial streak (Fig. 2.18), corrected with respect to the diffraction angle θ. The scattering intensity decreases monotonically up to a spinning speed of 5000 m/min, while a shoulder, or maximum, is observed in the scattering curve at spinning speeds higher than 6000 m/min. Since a broad peak appears in the scattering profile at spinning speeds of 8000–9000 m/min, an

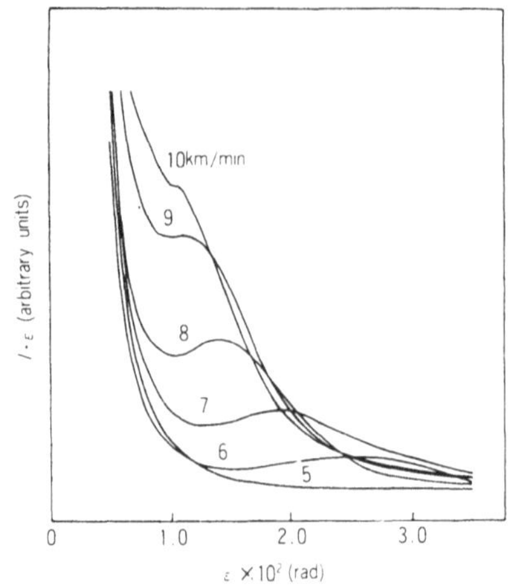

2.18 Equatorial intensity of SAXS from PET filaments spun at various spinning speeds.[5,7]

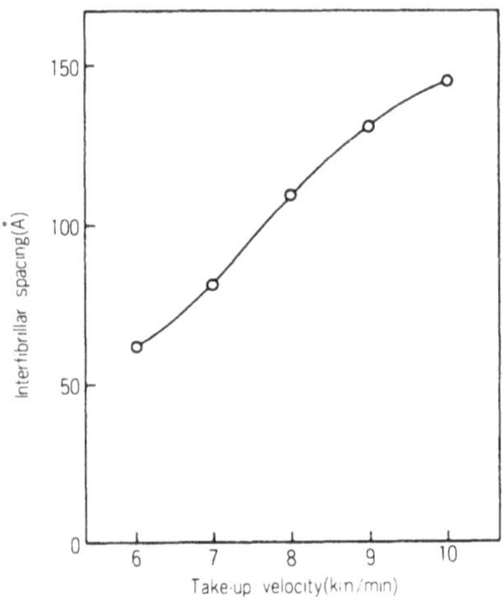

2.19 Interfibrillar distance as a function of spinning speed.[5,7]

approximate inter-microfibrillar distance was estimated by applying Bragg's equation. It was found to increase with the spinning speed as shown in Fig. 2.19. The inter-microfibrillar distance was estimated to be 62 Å at a spinning speed of 6000 m/min, and to increase to 145 Å at 10 000 m/min. The equatorial streak and the intensity in the meridional direction also increase with the spinning speed. Taking into account these results, the authors proposed a fine structure model for the skin part as shown schematically in Fig. 2.20.

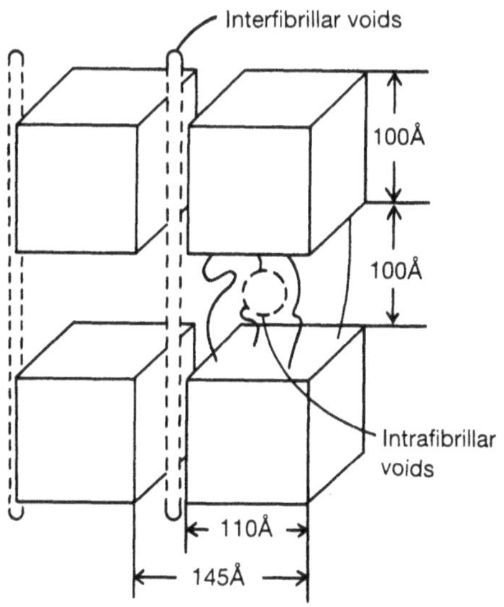

2.20 Fine structure model for spinning speed 10 000 m/min (from Ishizaki *et al.*[10]).

2.2.6 Oriented meso-phase

The fine structure, or high-order structure, of fiber has been analyzed in terms of a two-phase model composed of crystalline and amorphous phases. In this analysis, all phases except the crystalline phase are assigned to the amorphous phase, where the amorphous phase is not well defined but represents a general non-crystalline phase. Shimizu *et al.* have introduced the concept of the oriented meso-phase in order to specify the structure of the non-crystalline phase. The crystallinity evaluated from the density hardly increases with the spinning speed up to 4000 m/min, although the orientation improves continuously as mentioned above. Thus the fiber structure is highly oriented but amorphous in the range of spinning speeds of 3000–4000 m/min. Wide-angle X-ray diffraction reveals only a circular halo from the as-spun yarn at spinning speeds of 1000–2000 m/min, where the intensity of diffracted X-rays is almost equal in the equatorial and meridional directions as shown in Fig. 2.21. When the spinning speed increases, the difference in diffracted intensity between the equatorial and the meridional directions increases, and becomes greatest at a spinning speed of 4000 m/min. Since the intensity in the equatorial direction has its maximum at $2\theta = 21°$, the molecular chain is supposed to assume a periodic structure of 4 Å. The diffracted X-ray intensity is separated into three parts ascribed to the crystalline phase, the oriented meso-phase and the amorphous phase as shown in Fig. 2.22. Each area is proportional to the fraction of the respective phase, and Fig. 2.23 shows the spinning speed dependence of the fraction of each phase.

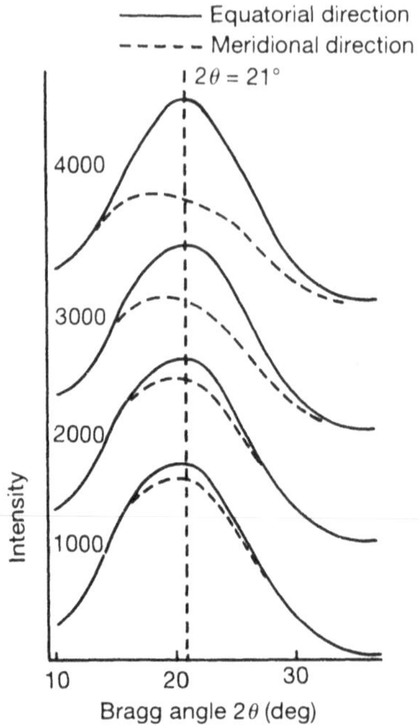

2.21 Equatorial and meridional intensity distribution of WAXS from amorphous region.[8,11,12]

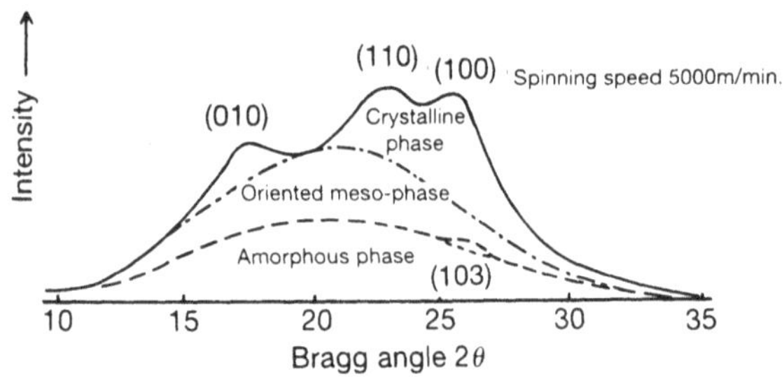

2.22 Separation of X-ray diffraction pattern into crystalline, oriented meso- and amorphous phases.[12]

Here the oriented meso-phase fraction increases and reaches its maximum of just over 20% at a spinning speed of 4000 m/min. When orientation-induced crystallization takes place (at a spinning speed of 5000 m/min), the oriented meso-phase transforms primarily into the crystalline phase.

The crystalline phase decreases and the amorphous phase increases at spinning speeds over 7000 m/min, while the oriented meso-phase exhibits a slight increase. This phase transformation is caused by the formation of an inhomogeneous structure in the radial direction and the consequent

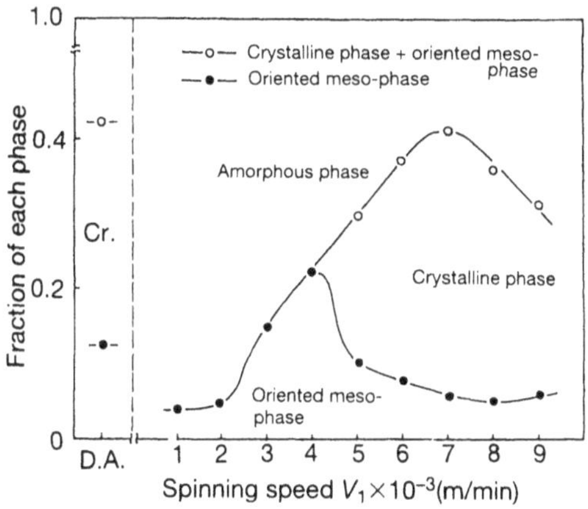

2.23 Fraction of crystalline, oriented meso- and amorphous phases as a function of spinning speed.[8,11,12]

deterioration of molecular orientation in the core at spinning speeds over 7000 m/min.

2.3 Structure formation in the high-speed spinning process

The structure and physical properties of high-speed-spun filament were reviewed in the preceding sections. This section describes how the fiber structure is formed during spinning.

2.3.1 Necking in the spinning process

Perez,[13] Matsui[14] and Shimizu *et al.*[8,15] confirmed the rapid thinning of filaments at one point in the spinline in the high-speed spinning of PET. This rapid thinning phenomenon is called 'necking' in the spinning process, since the phenomenon resembles necking at drawing. The filament diameter of the spinline was measured as a function of the distance from the spinneret as shown in Fig. 2.24. (See reference 8 for details of the online measuring system for filament diameter.)

The filament diameter decreases monotonically from the spinneret to the take-up winder at a spinning speed of 4000 m/min. When the spinning speed exceeds 6000 m/min, rapid thinning of filament diameter i.e. necking, is observed at a certain point, although gradual thinning of the diameter is observed for about 50 cm below the spinneret regardless of the spinning speed. The take-up filament diameter is almost equal to that at the completion of necking, and the necking onset shifts up-stream in the spinline with increasing spinning speed. That is, larger and more rapid deformation will take place in filaments at a higher spinning speed.

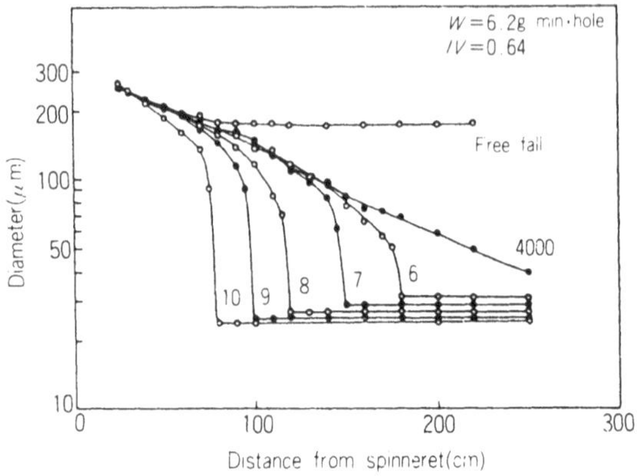

2.24 Change of filament diameter in spinline at 4000–10 000 m/min.[16]

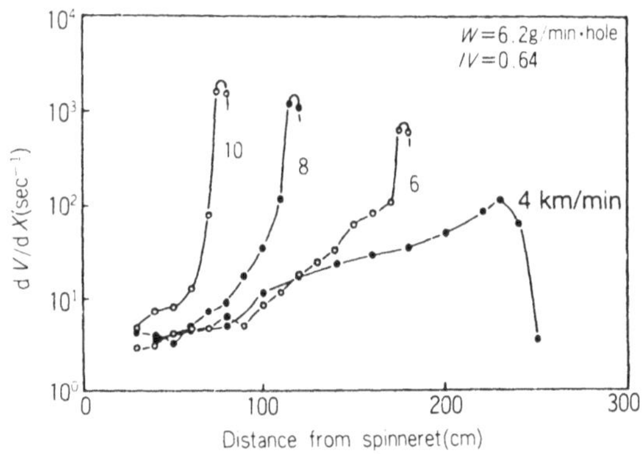

2.25 Change of deformation rate in spinline.

Figure 2.25 shows values of the deformation rate dv/dx averaged over 5 cm intervals from the data presented in Fig. 2.24. The deformation rate reaches a maximum at the necking point and the maximum value of the deformation rate is greater at a higher spinning speed. For example, the maximum deformation rate is 100 sec^{-1} at a spinning speed of 4000 m/min, where no necking is observed, whereas the rate becomes over 1500 sec^{-1} at a spinning speed of 10 000 m/min. Since the deformation rate estimated here is an average value over 5 cm intervals, the deformation rate may exceed 20 000 sec^{-1} at the necking spot. High-speed spinning is the only known example of this extremely large deformation rate in the drawing process.

The spinning stress was measured in the thinning process, and the spinning stress along the spinline was evaluated according to the following equation:[16,17]

$$F = F_L + \int_x^L \rho A g \mathrm{d}x - W(V_L - v) - \int_x^L C_f(\rho_a/2)v^2 \pi D \mathrm{d}x \quad [2.3]$$

$$C_f = 0.37 Re^{-0.61} \quad [2.4]$$

Here x denotes the distance between the spinneret and a point on the spinline, L the distance between the spinneret and the take-up point, F the spinning stress at a point x, F_L the take-up stress, ρ the polymer density, A the cross-sectional area of a filament at x, g the gravitational acceleration, W the polymer output, V_L the take-up velocity, v the filament velocity at x, ρ_a the air density, C_f the air drag coefficient, D the filament diameter at x, and Re the Reynolds number.

Equation 2.3 contains a second term due to the polymer mass, a third term due to the inertial force and a fourth term due to the air resistance. The surface tension was neglected and Matsui's equation[14] was adapted for C_f. Thus the spinning stress at a certain point x is given by subtracting the inertial force generated between x and the take-up point and the stress gained by air resistance from the sum of the take-up stress at the bottom of the spinline and the polymer mass at x. The results are shown in Fig. 2.26. The spinning stress increases abruptly at the onset of necking, which takes place at a lower spinning stress on increasing the take-up velocity (spinning speed). In particular, the spinning stress increases from 10^5 dyne/cm^2 to the order of 10^8 dyne/cm^2 at a spinning speed of 10 000 m/min. This abrupt increment of the spinning stress is observed at a point which shifts upstream in the spinline and is about 40 cm below the spinneret at a spinning speed of 10 000 m/min. The analysis indicates that the spinning stress is primarily determined by the inertial force in the upstream part of the spinline during ultra-high-speed spinning.

The elongational viscosity can be evaluated from the deformation rate and spinning stress according to the rheology equation:

$$\sigma = \lambda(\mathrm{d}v/\mathrm{d}x) \quad [2.5]$$

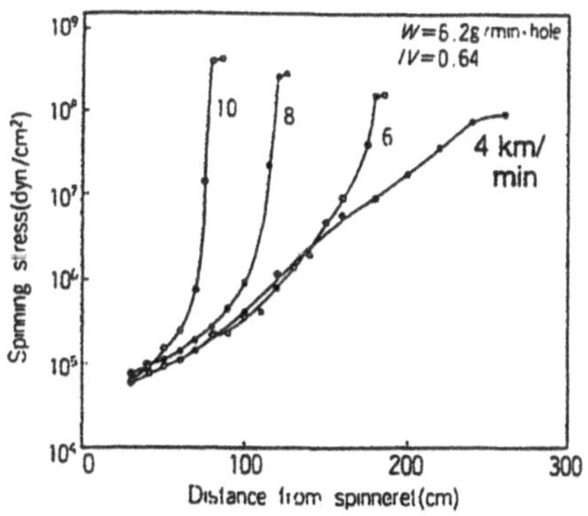

2.26 Change of spinning stress in spinline.[16]

Here σ denotes the spinning stress, λ the elongational viscosity, x the distance from the spinneret, and v the filament velocity at a point x.

The elongational viscosity is plotted against the distance from the spinneret in Fig. 2.27. Here a monotonic increase was observed for the elongational viscosity as the distance x increased at a spinning speed of 4000 m/min., when no necking took place. When necking takes place, at spinning speeds exceeding 6000 m/min, the elongational viscosity increases up to a certain distance, and then drops locally. The point where the elongational viscosity drops shifts upstream in the spinline and the elongational viscosity decreases with increasing spinning speed. After the localised drop, the elongational viscosity again increases steeply and the filament thinning is completed.

This localised drop in viscosity is supposed to correspond with the start of necking. Table 2.3 summarizes the filament velocity, the spinning stress, the elongational viscosity and the filament temperature at the start of necking. At the limiting spinning speed for necking (6000 m/min), the highest filament velocity (over 2000 m/min) was observed at the starting point of necking, while the filament temperature was low (125°C) but the spinning stress was high (1.8×10^7 dyne/cm^2). Shimizu *et al.*[8,11] and Kikutani[12] speculated about this behavior as follows. The chain orientation and crystallinity will be improved considerably by necking. The chain orientation and crystallization should take place within a short time since in that region the filament velocity is as high as 3000 m/min at a spinning speed of 5000 m/min. When the chain orientation improves, PET forms a pre-crystalline structure, which promotes easy filament drawing and eventually the velocity gradient increases. Accordingly the chain orientation further improves, the crystallization is accelerated, and

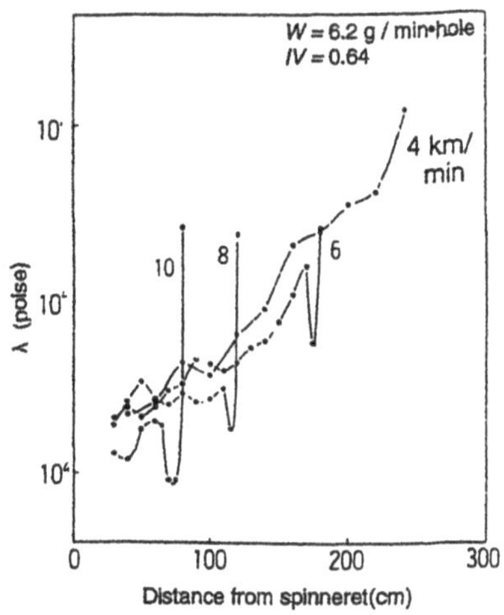

2.27 Change of elongation viscosity in spinline.[16]

Table 2.3 Filament velocity, deformation rate, spinning stress, elongational viscosity and filament temperature at the onset of necking

Take-up velocity (m/min)	Fiber velocity (m/min)	dv/dx (sec^{-1})	σ (dyn/cm^2)	λ (poise)	Temp. (°C)
6000	2014	111	1.8×10^7	1.6×10^5	125
7000	928	164	5.2×10^5	3.2×10^4	150
8000	887	119	3.7×10^7	3.1×10^4	180
9000	493	65	1.1×10^7	1.7×10^4	199
10000	288	18	3.5×10^7	1.9×10^4	221

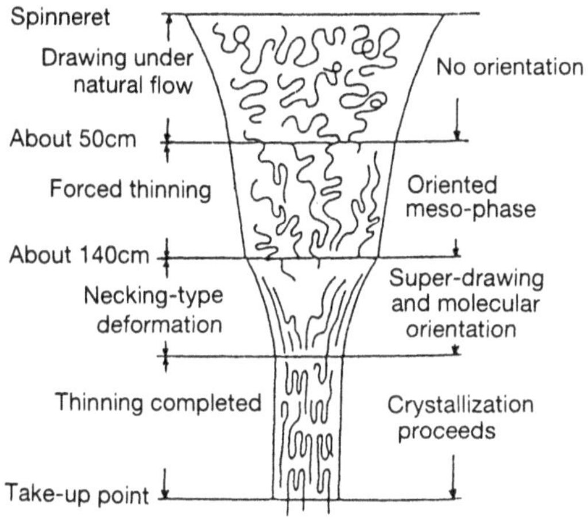

2.28 Change of fiber structure in high-speed spinning.[8,11]

the oriented meso-phase is formed. Then the elongational viscosity increases due to crystallization, and finally the system is fixed. This situation is shown schematically in Fig. 2.28.

In the case of ultra-high-speed spinning at 10 000 m/min, the filament exhibited a low velocity (288 m/min), a high temperature (221°C) and an extremely low spinning stress (3.5×10^5 dyne/cm^2) at the starting point of necking. This result suggests that chain orientation and subsequent crystallization are minimal at the starting point of necking in the case of the ultra-high-speed spinning at over 8000 m/min. The change in elongational viscosity at necking is supposed to be caused by a structural change resulting from the competition between the intermolecular interaction in the polymer melt and the spinning stress. Thus necking does not necessarily accompany orientation-induced crystallization, but rather takes place without chain orientation and crystallization.

Although this theory has stimulated discussion of the mechanism of formation of the fiber structure, no experimental evidence has yet been given to verify the speculation.

2.3.2 Fiber structure formation in the spinline

At which point in the spinline will the fiber structure be formed in the case of ultra-high-speed spinning such as at 10 000 m/min? Figure 2.29 shows the filament temperature in the spinline measured by a non-contact type infra-red radiation thermometer. The filament temperature decreases linearly down to a point 100 cm below the spinneret at spinning speeds of 4000 m/min and 8000 m/min. The filament temperature further decreases monotonically almost to the take-up in the case of a speed of 4000 m/min. At a spinning speed of 8000 m/min, the filament temperature drops abruptly by thinning when the necking takes place. The temperature drop becomes slower when necking is completed and then a temperature rise is observed for 20 cm downstream. In Fig. 2.30, the observed cooling curve is compared with the calculated cooling curve for a spinning speed of 8000 m/min. The filament temperature was calculated from the following equation by assuming that no

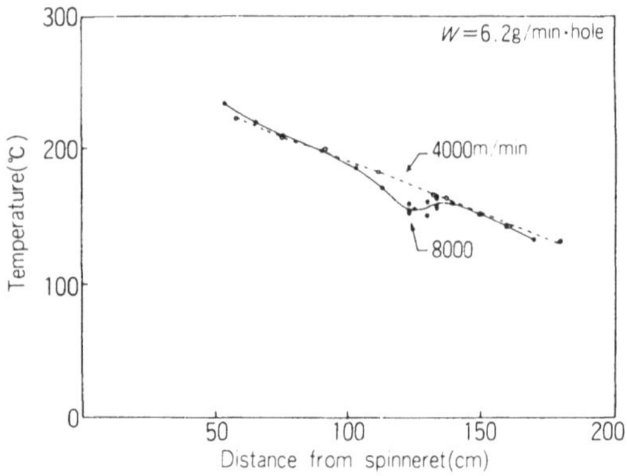

2.29 Change of filament temperature in spinline at 4000 m/min and 8000 m/min.[17]

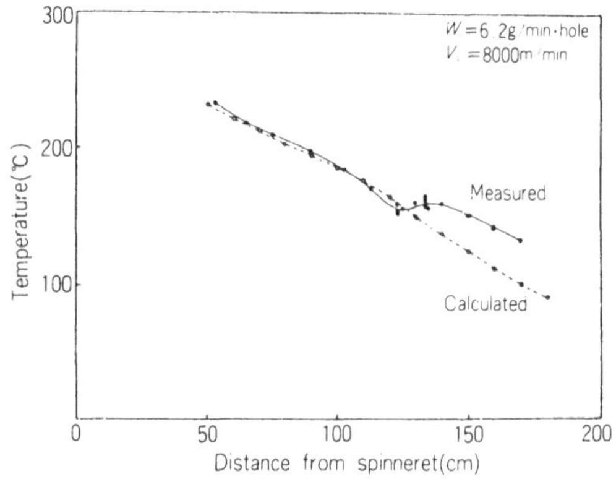

2.30 Calculated and measured filament temperatures in spinline.[17]

crystallization would take place. The ambient temperature was set to 150°C for 10 cm below the spinneret, and 20°C elsewhere.

$$Nu = hD/k_a = 0.42Re^{0.334}[1+(8v_y/v)^2]^{0.167} \quad [2.6]$$

$$WC_p(\mathrm{d}T/\mathrm{d}x) = -\pi Dh(T-T_a) \quad [2.7]$$

Here W denotes the polymer output, C_p the specific heat of polymer, T the filament temperature, T_a the ambient temperature, D the filament diameter, Nu the Nusselt number, k_a the thermal conductivity, h the heat transfer coefficient, Re the Reynolds number, v the filament velocity and v_y the velocity of cooling air ventilated vertically.

The calculated temperature agreed well with the observed filament temperature until necking was completed, as seen from Fig. 2.30. The observed filament temperature deviated from the calculated value downstream from this point. This discrepancy is probably due to crystallization which is not considered in the model calculation. The exotherm at crystallization caused a slight temperature rise by compensating for the cooling of the filament.

Figures 2.31 and 2.32 show the changes of the filament diameter, the spinning stress and the birefringence Δn in the spinline at spinning speeds of 4000 m/min and 8000 m/min. At a spinning speed of 4000 m/min, Δn increases gradually from the point where the filament diameter and the filament velocity are 60 μm and 2000 m/min, respectively, and saturates at the completion of filament thinning.

At a spinning speed of 8000 m/min, Δn increases rapidly from the spot where the necking terminates. The filament diameter and the filament velocity at this spot are 35 μm and 5500 m/min, respectively. The spinning stress at this spot is 10^8 dyne/cm^2, which is enough to promote chain orientation. Δn then increases from 0.04 (at the spot where the necking terminates) to 0.11 for 15 cm downstream. The filament

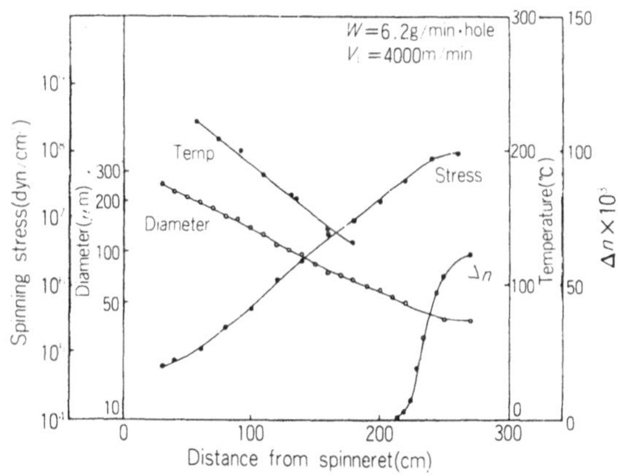

2.31 Change of characteristic properties in spinline at 4000 m/min.[17]

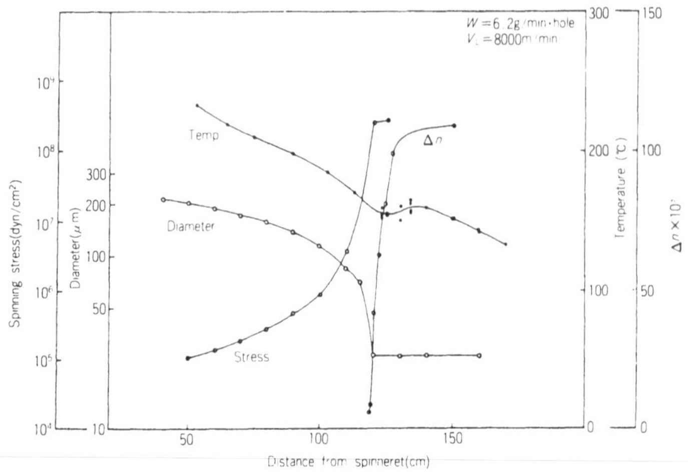

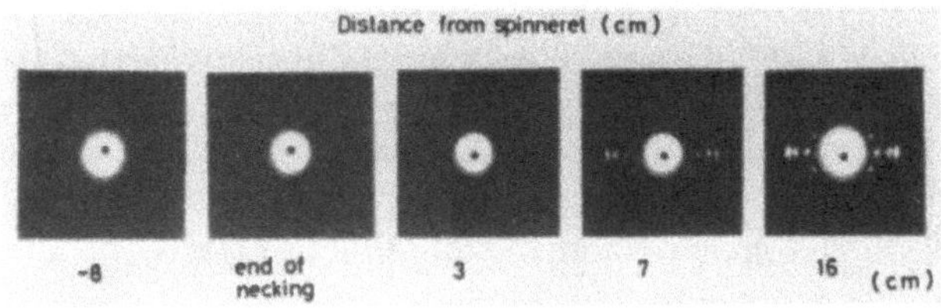

2.32 Change of characteristic properties in spinline at 8000 m/min.[17]

temperature rises in accordance with this rapid increase of Δn, and crystallization proceeds.

The wide-angle X-ray diffraction pattern reveals only an amorphous halo from the filament sampled at a point 8 cm upstream from the neck. The small spots from crystallites appear first at the completion of necking. Distinct crystallite spots are observed from the filament sampled at a point 7 cm downstream from the end of necking, and the crystallites are well grown at a point 16 cm downstream from the end of necking. The results indicate that the crystallization starts at the end of necking and progresses rapidly for 20 cm downstream. This means that a high spinning stress is exerted to promote rapid orientation-induced crystallization and the fiber structure is established within a short period after the necking is completed.

This thinning process and the subsequent structure formation are schematically shown by Ishizaki *et al.*[10] and Ichara *et al.*[5] as in Fig. 2.33. When a polymer melt is extruded from the spinneret at a spinning speed of 10 000 m/min, it yields to decrease the elongational viscosity and necking develops rapidly at a slow filament velocity of 300 m/min because the balance between the cohesive force and the spinning stress is

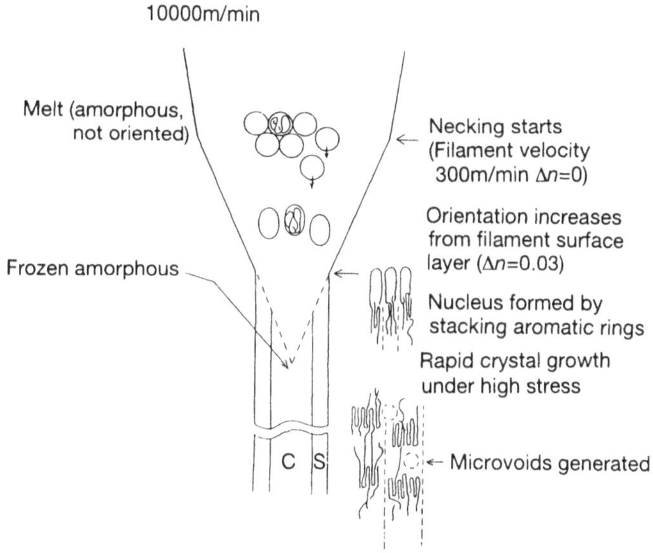

2.33 Model of fiber structure formation (from Iohara *et al.*[5] and Ishizaki *et al.*[10]

disturbed. Molecular orientation progresses mainly at the filament surface, and Δn increases to 0.03 with the completion of necking. Then the aromatic rings become stacked to form crystallite nuclei, and the crystallites grow rapidly for several cm downstream. Although this process penetrates gradually from the yarn surface to the inside, the crystal growth is so rapid that many microvoids are generated in the non-crystallized area such as between and within fibrils. Since the spinning stress concentrates on the surface, the molecular orientation will not improve in the central part of filament where the amorphous structure is frozen-in. The progress of this structure formation is shown schematically in Fig. 2.33.

2.3.3 Influence of air resistance in a high-speed spinning process

The spinning tension in high-speed spinning consists of the inertial force and the air resistance. It is generally accepted that structure formation is determined mainly by the inertial force and that the air resistance plays no significant role in ultra-high-speed spinning at over 8000 m/min. However, it is also reported that the air resistance may influence the formation of fiber structure in fine filaments (less than 1 denier) in high-speed spinning at 4000–5000 m/min. The influence of air resistance in the spinning process is examined in this section. Figure 2.34 shows the spinning test procedure where the air resistance exerted on the filaments is varied by changing the distance between the spinneret and the take-up

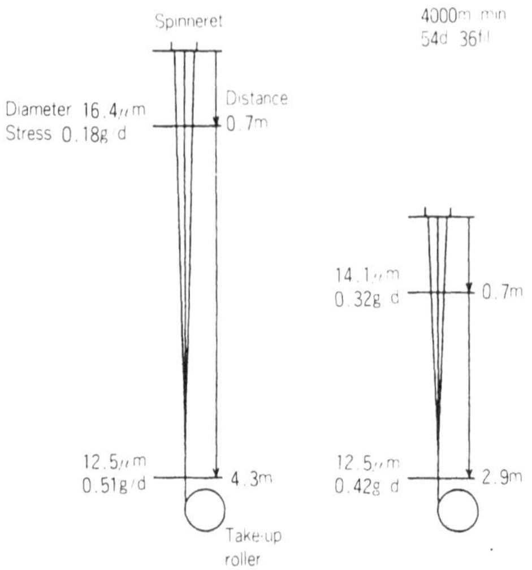

2.34 Schematic view of spinning test precedure for changing filament running length.[18]

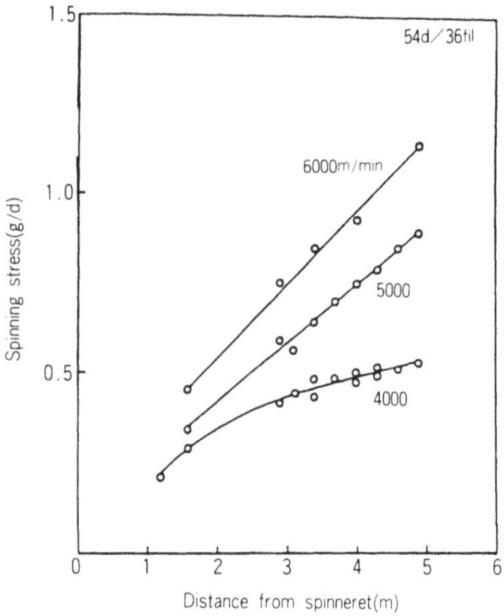

2.35 Spinning stress at take-up point as a function of running length at various spinning speeds.

roller while the take-up speed, the polymer output and the cooling condition are kept unchanged.

The spinning stress (the take-up stress) changes at the take-up point as shown in Fig. 2.35 when the filament running length is varied. Here the linear density of multi-filament yarn (36 filaments) is 54 d. The take-up stress increases monotonically as the running length of filament increases at any spinning speed from 4000 m/min to 6000 m/min. The structure and physical properties are practically the same for the filaments spun at 4000–6000 m/min. For example, at a spinning speed of 4000 m/min, the birefringence Δn and boil-off shrinkage are measured as a function of spinning stress (Fig. 2.36) where the filament linear density and the spinning speed are fixed at 1.5 d and 4000 m/min. When the takeup stress exceeds 0.45 g/d, both Δn and the boil-off shrinkage exhibit a sudden increase. Since the take-up speed, the polymer output and the cooling condition are kept unchanged, the large change observed in Δn and the boil-off shrinkage is considered to be due to the influence of air resistance. The equatorial X-ray diffraction intensity from the same sample exhibits crystallite diffraction at a low takeup stress (0.21 g/d), but no crystallite diffraction pattern is observed when the takeup stress exceeds 0.45 g/d (Fig. 2.37). Thus the fiber assumes an amorphous and highly oriented structure when the spinning length of filament becomes longer and the take-up stress exceeds a certain limiting value, while all other spinning conditions are kept identical. This suggests that the deterioration of crystallinity is due not to the higher cooling rate for the thinner filament but to the air resistance influencing the structure formation in the downstream region of the spinline.

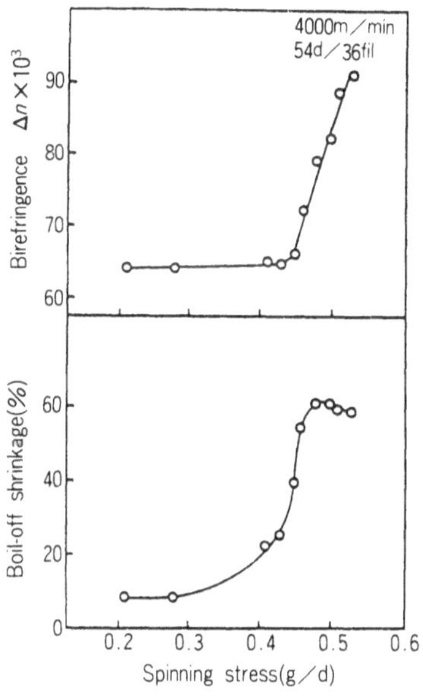

2.36 Δn and boil-off shrinkage as functions of spinning stress at 4000 m/min.[18]

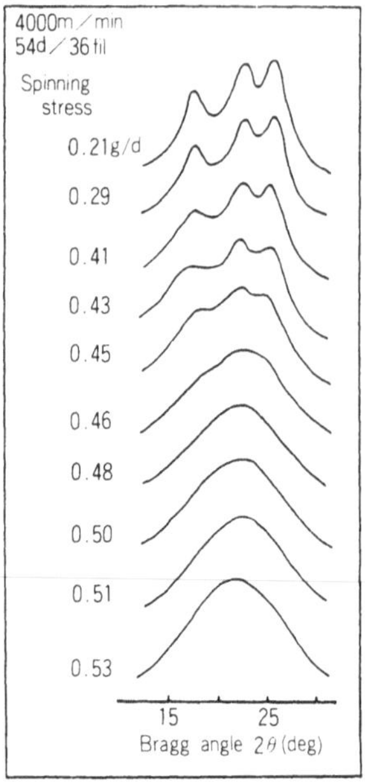

2.37 Equatorial X-ray diffraction profile as a function of spinning stress at take-up point (4000 m/min spinning speed).[18]

In Fig. 2.38, the filament velocity is estimated from the filament diameters and the spinning stresses at a point 70 cm below the spinneret and at the take-up point. At this point the running lengths of filament are 2.9 m and 4.3 m with a constant filament count of 1.5 d at a constant spinning speed of 4000 m/min. Figure 2.38 suggests that the filament

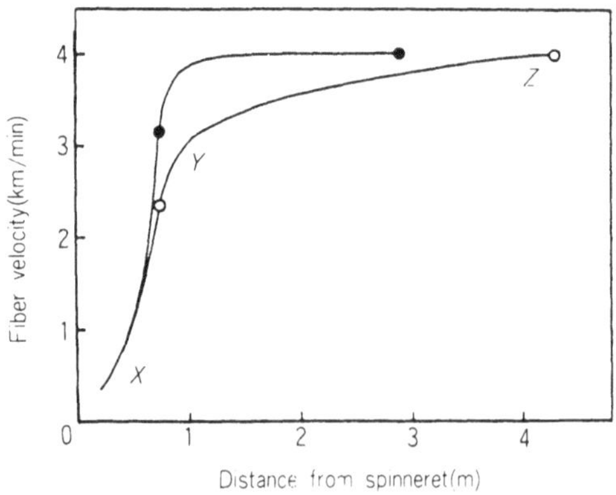

2.38 Estimated filament velocity for two different running lengths (4000 m/min spinning speed).[17] (The open circles represent a spinneret take-up distance of 430 cm; the filled circles represent 290 cm.)

velocity increases in two steps when the running length of filament is long (4.3 m). The first step from X to Y in Fig. 2.38 corresponds to the thinning process in the conventional melt spinning, and the second step from Y to Z is considered to be deformation by cold drawing since the filament temperature is rather low in this region. When the running length is long and the air resistance is high, the deformation by cold drawing in the second step reduces the filament velocity at Y to a lower value than at the short running length, and consequently crystallite growth is suppressed. If the running length of filament becomes longer, the ratio of cold drawing increases, the filament velocity is further reduced and no crystallization takes place at Y. In this case, molecular chains are highly oriented by cold drawing in the second step, and the fiber assumes a highly oriented amorphous structure.

Cold drawing depends upon a subtle balance of the air resistance and the crystallization rate. In high speed spinning at 4000–5000 m/min, the air resistance is large and the crystallization rate is low, so that the filament possesses less crystalline structure and cold drawing results.

2.3.4 Formation of skin–core structure

The interference pattern of refraction changes during necking. Figure 2.39 shows the interference pattern from the 6000 m/min spun filament sampled during necking. The pattern change starts from the surface layer of the filament, and develops in the filament core going downstream in the spinline. This suggests that the stress concentration caused by cooling promotes molecular orientation first in the surface layer and then in the filament core. The molecular orientation is homogeneous in the filament cross-section at a point 7 cm below the end of necking in the case of spinning at 6000 m/min. However, in the case of ultra-high-speed spinning at over 8000 m/min, the stress is mainly concentrated in the

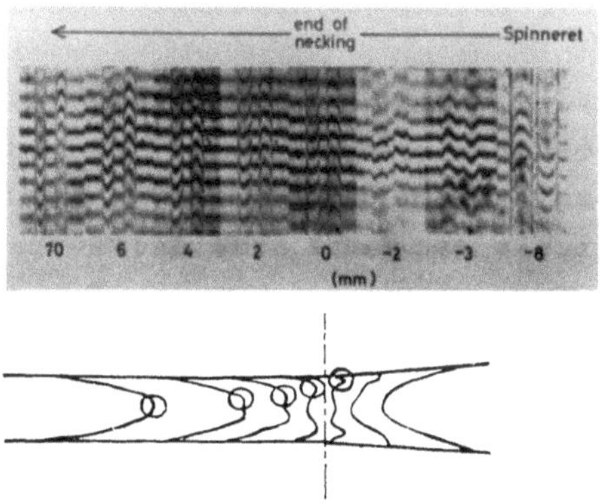

2.39 Interference refraction patterns of filaments sampled from spinline.[5,10,17]

surface layer, and the amorphous structure is frozen in the filament core before becoming highly oriented. In consequence, the skin–core structure is formed.

2.4 Applications of high-speed spinning

2.4.1 Applications specific to high-speed-spun yarn[5,19]

High-speed spinning is widely employed as a very efficient production system. There have been some attempts to utilize the characteristics of high-speed-spun yarn to develop yarns with new properties.

Asahi Chemical Co. developed an easily-dyeable polyester yarn which can be dyed at normal pressure.[20] When combined with specific cooling conditions, high-speed spinning at over 7000 m/min yields a polyester yarn which can be dyed under normal pressure and possesses the structure and properties listed in Table 2.4. The yarn is characterized by low-density amorphous regions and a high crystallinity. The low density improves the dyeability, and the high crystallinity assures thermal stability. The temperature dispersion curve of dynamic viscoelasticity at 110 Hz confirms the low density and the small fraction of amorphous regions from its tan δ peak temperature (equivalent to the α dispersion), T_{max} shifting to a lower temperature and its peak value $(\tan\delta)_{max}$ being small. The crystallinity evaluated by wide-angle X-ray diffraction is as high as 70%. The crystallites are large in size as evaluated from the (110) reflection, and are highly oriented as deduced from the (010) reflection, so that large and stable crystallites are formed. Δn = 0.03–0.11 which is lower than the 0.173 obtained for typical FOY, suggesting lower orientation in the amorphous region. The thermal stability is good as shown by a high retention ratio (E'_{220}/E'_{150}) of the storage modulus at 220°C to 150°C, which was found to be over 0.75, compared with 0.53 for FOY.

Another characteristic of the yarn is a specific inhomogeneity of the fiber structure in the radial direction within filaments. This inhomogeneity is measured by the difference of refractive index $\Delta n(0.8-0)$ defined as $\Delta n\ (0.8-0) \equiv n_{0.8}-n_0$, where n_0 denotes the average refractive index at the filament center for the light oscillating in parallel to the fiber axis and $n_{0.8}$ denotes the refractive index at a point $0.8r$ from the centre, where r is the filament radius. $\Delta n(0.8-0)$ is found to be in the range 0.01 to 0.08, in contrast to 0.002 observed for FOY. Here the distribution of the refractive index is symmetrical with respect to the filament centre.

A spinning speed of over 7000 m/min is a prerequisite to produce the yarns possessing the characteristic structure above. The spinning threadline is provided with a heating zone attached to the bottom of the spinneret and a fluid suction device further downstream in order to

Table 2.4 Characteristics of normal-pressure dyeable yarn

		1	2	Reference yarn 1	FOY	Desirable range
Spinning						
Take-up velocity	(m/min)	9000	7000	6000	1500	7000 m/min
Draw ratio					3.3	no drawing
Viscoelasticity						
T_{max}	(°C)	102	104	114	132	85–110°C
$(\tan \sigma)_{max}$	(–)	0.126	0.129	0.182	0.110	0.115–0.135
E'_{30}	(g/d)	74	73	49	97	over 50 g/d
Refractive index						
Δn	(–)	0.083	0.108	0.116	0.173	0.03–0.110
Average refractive index	(–)	1.660	1.677	1.673	1.691	1.65–1.70
Difference of refractive indices $n_{0.8}-n_0$	(–)	0.031	0.026	0.009	0.002	0.01–0.08
X-ray						
Degree of crystallinity	(%)	77	75	66	58	over 70%
Crystallite size	(Å)	65	56	36	18	over 55Å
Degree of orientation	(%)	94	94	91	90	over 90%
Yarn property						
Tenacity	(g/d)	3.5	4.3	4.1	5.0	over 3 g/d
Elongation at break	(%)	25	46	62	22	20–60%
Boil-off shrinkage	(%)	2.9	3.2	3.9	7.8	
E'_{220}/E'_{150}		0.79	0.70	0.64	0.53	over 0.75
Equilibrium exhaustion	(%)	89	75	52	49	

prevent the filaments from being quenched or cooled inhomogeneously (see Fig. 2.40).

The equilibrium exhaustion with disperse dye at 100°C is as high as 75–89% (cf. 49% for FOY dyed under the same conditions), which is similar to the 85% observed for FOY high-pressure dyed at 130°C.

2.4.2 Improvement of physical properties of yarn

It has long been known that the mechanical properties of a filament can be improved by heat-treatment of running filament in the spinline (spinline heating: SLH). For example, Griehl and Versäumer[21] found deterioration of elongation of nylon 6 in 1958 when a heating zone was provided round the spinline during spinning. In 1969 Hamana[22] reported that the filament birefringence Δn was increased from 0.0064 to 0.0104 by providing a 30 cm long heating jacket at 200°C round the spinline 70 cm below the spinneret when spinning at 900 m/min. Since the spinning speed here is low, the value of Δn reached is still low and the resulting filaments cannot be used in practice as drawn filaments.

This idea of spinline heating was then combined with high-speed spinning, and many patents were applied for covering the one-step

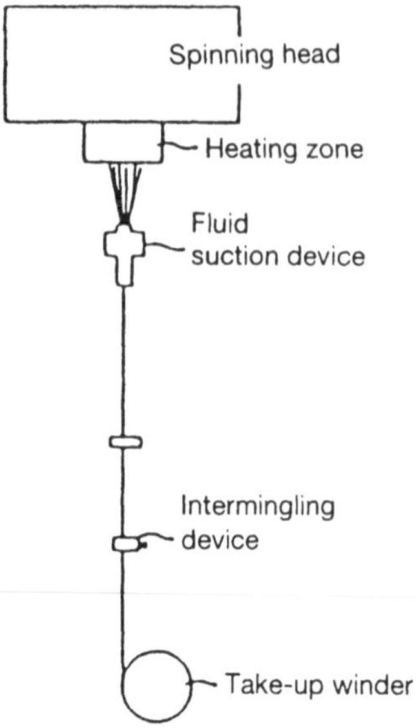

2.40 Schematic view of spinning.

production of FOY-level yarn at spinning speeds of 4000–6000 m/min at the time when no high-speed take-up winders were available. In this range of spinning speeds, the mechanical properties of yarn are not normally satisfactory in terms of the tensile strength (4.5 g/d), the elongation-at-break and the yield point. Various SLH methods were developed as seen from the patents discussed below.

Toyobo[23] succeeded in producing polyester yarn of tensile strength 4.5 g/d, an elongation at break of 35% and a dry-heat shrinkage at 160°C of 12% with a spinning speed of 4000 m/min by providing a 2 m long heating jacket (200°C) about 30 cm below the spinneret (see Table 2.5). When the spinning speed was increased to 6000 m/min, the tensile strength improved to 5.0 g/d. The position of the heating zone is an important factor in bringing the SLH effect into full play. It should be adjusted to a position where the filament is thinned by 90–100% of the amount required to reach the denier of the take-up filament.

Toray[24] employed pressurized vapor for SLH to re-heat cooled filaments for drawing. Since high filament tension is required for better drawing, the filament is run through a heating chamber filled with pressurized vapor to take advantage of the increased air resistance exerted by pressurized vapor of higher density. The entrance and exit of the heating chamber should be narrow enough to pass the filaments and sealed aerodynamically to retain the pressure. The heating chamber cuts the associated air flow at the narrow entrance and exit, and the effect of heat treatment is improved. The heating chamber is made of a cylindrical vessel 40 cm in length and 60 mm in inner diameter with slits of

Table 2.5 Effect of SLH (Toyobo[23])

	Spinning speed (m/min)	SLH	Tenacity (g/d)	Elongation (%)	Dry shrinkage at 160°C (%)
1	4000	none	2.7	85	22.1
2	4000	heating jacket	4.5	35	12.0
3	6000	none	3.3	68	5.7
4	6000	heating jacket	5.0	29	9.2

Table 2.6 Effect of SLH (Toray[24])

	Vapor pressure (kg/cm^2)	Tenacity (g/d)	Elongation (%)	Boil-off shrinkage (%)
1	(none)	3.5	82	11.5
2	1.0	4.0	33	9.0
3	3.2	4.4	23	4.9

dimensions 0.2 mm (W) × 0.2 mm (D) × 30 mm (L) provided on the top and bottom of the cylinder. The filament linear density produced is 1 d, and its tensile strength is improved to 4.4 g/d (see Table 2.6).

Toray[25] combined a conventional drawing and heat-treatment process with SLH to improve the tensile strength of polyester yarn. Here, polyester yarn of high strength is obtained by heat treating SLH-spun yarn under a tension ratio of 0.8–1.1 at 80°C. Teijin[26] converted the takeup roller for SLH-spun yarn into a tapered heating roller to draw by 1–20%. The SLH method is also applied to balance the tensile strength and the elongation-at-break of yarn, to produce high-speed-spun yarn with the mechanical properties of low-speed-spun yarn, and to improve the efficiency of high-speed spinning processes.

Some attempts have been made to cool filaments in the spinline rather than to heat them. According to Kikutani,[27] the aim of cooling is threefold:

1 To produce highly oriented amorphous yarn by quenching just after necking.
2 To produce highly oriented amorphous yarn by controlling the cooling water temperature and deforming the yarn in water.
3 To produce highly oriented crystalline yarn by applying boiling water as the cooling medium.

As an the example of point 1 (above), Du Pont[28] has patented the use of high-speed-spun yarn as an as-spun yarn for false twisting. The crimping characteristics deteriorate when crystalline yarn is processed for false twisting. As-spun POY made at a spinning speed below 4000 m/min is employed conventionally as an as-spun yarn for false twisting. Highly oriented amorphous yarn can be produced by water-cooling high-speed-spun polyester yarn in the spinline at spinning speeds of over 5500 m/min according to the Du Pont patent. Table 2.7 shows an example where 75 d/17 filament polyester is spun at a spinning speed of 5669 m/min. Highly oriented amorphous yarn with high boil-off shrinkage is obtained by quenching at a point 96.5 cm below the spinneret. Here the quenching position plays a key role. For a filament linear density of 4.4 d at a spinning speed of 5994 m/min, amorphous yarn with 56% boil-off shrinkage is obtained by quenching at a point 76 cm below the spinneret, whereas considerable crystallization occurs on shifting the quenching position to a point 81 cm below the spinneret as indicated by a low boil-off shrinkage of 10.8%.

Toyobo applied the cooling method of point 2 (above) to producing high-tenacity PET yarn.[29] Highly-oriented amorphous yarn is required for as-spun yarn to process into high-tenacity PET yarn. According to this patent, a highly-oriented amorphous structure is formed by cooling filaments in the spinline during high-speed spinning. In an example at a spinning speed of 4200 m/min (Table 2.8), water cooling is introduced at a point less than 5 cm upstream towards the spinneret from the point where the filament thinning is completed by air cooling. By air cooling, the filament thinning is completed at a point 48 cm below the spinneret, so the filaments are quenched in a water bath 25 cm long from a point 36 cm below the spinneret in the case presented in Table 2.8. Highly oriented amorphous as-spun yarn is obtained, and by direct drawing the tenacity and modulus reach the high values of 11.3 g/d and 171 g/d, respectively.

Although the cooling bath temperature is rather high and thus the process is not well distinguished from SLH in the case of point 3 (above),

Table 2.7 Water-quenching and boil-off shrinkage

Quench bath position (cm below spinneret)	Boil-off shrinkage (%)	Density (g/cm^3)	X-ray diffraction
96.5	67	1.357	amorphous halo
107	15	1.385	crystalline spots

Table 2.8 Cooling system and yarn properties

	Undrawn		Drawn		
Cooling system below spinneret	Δn	Density (g/cm^3)	Tenacity (g/d)	Elongation (%)	Modulus (g/d)
Water bath (36cm)	0.150	1.3582	11.3	7.4	171
Air cooling	0.088	1.3722	8.2	7.6	148

a report by Cuculo *et al.*[30] can be cited as an example. One-step high-speed spinning yields yarn with a tenacity of, at most, 4–5 g/d. Cuculo *et al.* obtained high-performance PET yarn for tire cords by controlling the spinline dynamically. They obtained a high-performance yarn of tenacity about 9 g/d and Δn over 0.21 from PET of IV 0.95 at a spinning speed of 4000 m/min by providing an LIB (liquid isothermal bath) of bath temperature 136°C below the spinneret (see Fig. 2.41).

2.4.3 Other applications

Although the ultra-high-speed spun yarn is not yet commercially utilized as feed stock for tire reinforcement, there have been several attempts to apply the high-speed spinning for the production of high-tenacity tire yarn of 9 g/d.

High-tenacity PET tire yarn has so far been produced by high-ratio drawing of low-speed-spun yarn of low crystallinity and low orientation, where the basic idea is represented by Du Pont's patent.[31] In 1987, Hoechst–Celanese[32] patented the production of high-tenacity yarn with good thermal stability and improved wear-resistance from highly oriented high-speed/high-tension spun yarn. Table 2.9 compares the two methods of Du Pont and Hoechst–Celanese. Here the spinning speed in the Hoechst–Celanese method is 1300 m/min, which is much higher

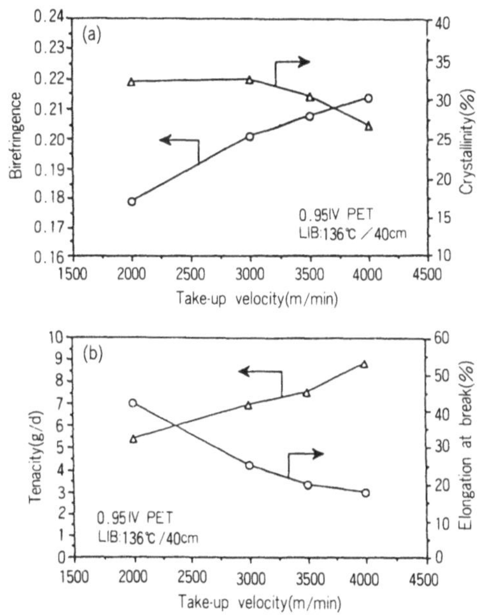

2.41 One-step spinning of high-performance yarn using LIB bath (from Cuculo *et al.*[30]). (a) Δn and (b) tenacity and elongation as functions of spinning speed.

Table 2.9 Tire yarn: spinning process, structure and properties

		Du Pont[31]	Hoechst–Celanese[32]
Spinning			
Atmosphere below spinneret		Heating shroud (375°C)	Heating zone (room temperature)
Cooling		[conventional]	homogeneous
Tension	(g/d)	0.0026	0.076
Spinning speed	(m/min)	228	1300
Draw ratio		6.25	2.52
Properties and structure			
UD filament Δn		0.0005	0.038
Tenacity	(g/d)	9.6	8.8
Elongation-at-break	(%)	15.6	6.8
Dry shrinkage	(%)	(14)	5.0
4.5kg load	(%)	(6.3)	6.1
Hysteresis loss	(in–lb)	(0.09)	0.014
Stability index		(0.8)	14.1
Tension index	(g/d)	(1250)	1148
Amorphous orientation		(0.6↑)	0.56
Long period	(Å)	180	140

Stability index = 1/(hysteresis loss × 175°C dry shrinkage).
Tension index = tenacity × modulus.
Bracketed figures indicate estimated values.

than the 228 m/min in the Du Pont method, although the spinning speed does not yet reach the usual POY range.

Tire yarn produced by the Hoechst–Celanese method exhibits lower heat-shrinkage, i.e. better thermal stability, resulting in a small hysteresis loss, less heat generation and improved wear-resistance. The tube-life (evaluated by the GY Malory Tube Test) of this tire yarn is 5–10 times longer and the temperature at the tube surface is 28°C lower than for the Du Pont low-speed-spun tire yarn according to the patent. Although the yarn strength is a little lower, the yarn is designed to have an excellent wear-resistance against the repeated elongation–compression load exerted by tire rotation.

These two spinning methods afford different fine fiber structures. The Hoechst–Celanese method yields a yarn with less oriented amorphous regions. Although this is a disadvantage from the viewpoint of tensile strength, the molecular chains in the amorphous region are less restricted, and in consequence heat generation is suppressed. In Fig. 2.42, the tube-life is plotted against the half-width of the α_a dispersion (the major dispersion in the temperature range 120–130°C in the viscoelastic temperature dispersion curve), which measures the orientation distribution of chain molecules in the amorphous region.[33] The tube-life improves as the half-width narrows, that is, the amorphous chain molecules orient more in one direction. Although the degree of amorphous orientation is low in the Hoechst–Celanese yarn, the orientation distribution is narrow and the two phases (amorphous and

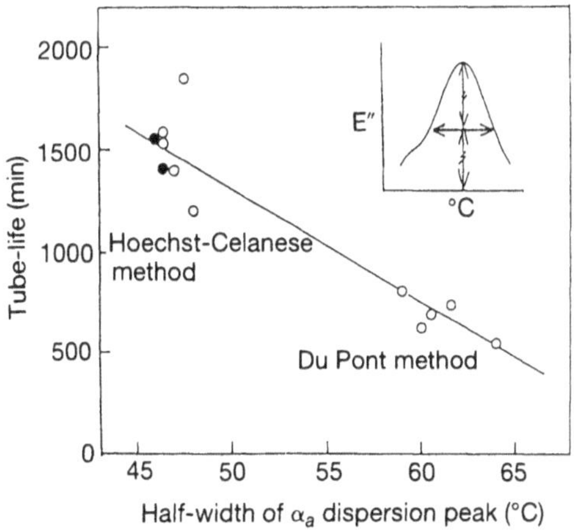

2.42 Tube-life and half-width of α_a-dispersion peak.[33]

crystalline phases) are finely mingled as indicated by a shorter long period (see Table 2.9). The wear-resistance improves accordingly.

The Hoechst–Celanese method has been modified and much improved to produce POY suitable for tire yarn. Commercial production of tire yarn by this method started in 1983 in Japan. Here the spinning speed is increased to the range of 2000–3000 m/min, and a higher molecular weight polymer is used to meet the mechanical and thermal requirements.

2.5 Present state of high-speed spinning operations

Asahi Chemical Ind. announced a plan to put high-speed spinning at over 6000 m/min into operation in June 1987. Production started in April 1988 with a production capacity of 5.1 t/day. The yarn (commercially termed TEC®) is spun at a spinning speed of 7000 m/min and is directly taken up on the roller without passing over a godet roller. Subsequently, Teijin, Toray and Mitsubishi Rayon announced plans to put their own high-speed spinning into operation as summarized in Table 2.10.

Asahi's high-speed spun yarn (TEC) is characterized by softness (low Young's modulus), easy dyeability and low thermal shrinkage in comparison with conventional FOY. The yarn is applied in linings for men's suits (softness required), tricot for car seats (easy dyeability required) and thin interlinings for garments (low thermal shrinkage required).

Teijin produces HOY (highly oriented yarn) by high-speed spinning at 6000 m/min through godet rolls. Their first plant produces speciality yarns such as anti-static, brilliantly dyeable, or cationic dyeable yarns. In their second plant, an FMS (flexible manufacturing system) is introduced to produce yarns from different polymers simultaneously, to produce

Table 2.10 High-speed spinning of PET in Japan

	Operational State (1993)			Remarks	Plan	
	Operation started (year-month)	Spinning speed (m/min)	Amount (t/month)		Fully-operational (year)	Amount (t/month)
Asahi Kasei	88–4	7000	300	Direct take-up, easy dyeability, soft hand	–	600
		5500	300	Godet		
Teijin	88–6	6000	1100	Godet, specific yarn, FMS	92	1100
Toray	90–3	6000	2060	Godet, versatile production unit	–	–
Mitsubishi Rayon	89–4	6000	500	Godet, provided with downstream processing system	91	720
Toyobo	?	6000	110	Godet (up to 4), also capable of spinning nylon	91	~450
Unitika	?	6000	?	Godet, for regular yarn	90	~800

yarns of different characteristics and to control each device independently by computer. The production system is also designed to meet a wide range of requirements such as rationalization by introducing a multi-position system, operational stabilization by introducing SHP (super high performance) or faster weaving with WJT (water jet looms).

Toray has developed a multi-purpose OSP (one-step process) ultra-high-speed spinning system, where the multi-filament yarn is spun at a spinning speed of 6000 m/min through godet rolls. Toray has also developed a 3 Kg take-up winder to cope with the production of many kinds of small lot. The total system is capable of producing a wide range of yarns including SILLOOK® and ultra-fine fiber, and also of easy dyeability, soft hand and good nap, and covers a whole series of processes from spinning to finishing.

2.6 Conclusion

This chapter describes high-speed spinning, especially of PET, as an example of the most advanced melt spinning technology. Although speciality yarns have been developed, ultra-high-speed spinning has a potential to produce higher-performance yarns which has not yet been fully exploited. Although PET Shingosen is highlighted in the market for its excellent touch superior to natural materials, it may be replaced with ultra-high-speed spun yarns, which can be produced more efficiently. A yarn with a higher-tenacity than the present tire yarn is also likely to be produced by ultra-high-speed spinning in the future. Ultra-high-speed spinning will advance further in terms not only of its efficiency, but in its potential to produce new types of high-performance yarns.

References

1 Du Pont, Jap. Pat. (Examined), 56-6,767, 56-6,768, 56-3,104.
2 Ueda A and Kanatsuna M, Reports of Research Institute of Textile and Polymers, MITI, 1971.
3 Shimizu J, Toriumi K and Tamai K, *Sen-i Gakkaishi*, **33**, T-208, 1977.
4 Ishizaki S, Proceedings of Kyoto International Symposium on Synthetic Fibres, 301, 1985.
5 Iohara K, Ohwaki S and Murase Y, Preprints of Gordon Conference, 1985.
6 Kamide K and Kurimoto T, *Sen-i Kikai Gakkaishi*, **38**, 268, 1985.
7 Fujimoto K, Thesis submitted to Tokyo Institute of Technology, 1989.
8 Shimizu J, *Sen-i Kikai Gakkaishi*, **38**, 243, 1985.
9 Shimizu J, Okui N and Kikutani T, *Sen-i Gakkaishi*, **37**, 135, 1981.
10 Ishizaki S, Iohara K and Fujimoto K, *Sen-i Gakkaishi*, **45**, 234, 1989.
11 Shimizu J, Kikutani T, Takaku A and Okui N, *Sen-i Gakkaishi*, **37**, T-135, 1981.
12 Kikutani T, Thesis submitted to Tokyo Institute of Technology, 1982.

13 Perez G and Leclise C, Proceedings of the 18th International Conference on Chemical Fibres (Dornbirn), 1979.
14 Matsui M, *Sen-i Gakkaishi*, **38**, 508, 1982.
15 Shimizu J, *Sen-i Gakkaishi*, **38**, 499, 1982.
16 Fujimoto K, Iohara K, Ohwaki S and Murase Y, *Sen-i Gakkaishi*, **44**, 53, 1988.
17 Fujimoto K, Iohara K, Ohwaki S and Murase Y, *Sen-i Gakkaishi*, **44**, 171, 1988.
18 Fujimoto K, Iohara K, Ohwaki S and Murase Y, *Sen-i Gakkaishi*, **44**, 477, 1988.
19 Murao Y, Preprints of the 8th Lecture on Fibre Material Research, Society of Polymer Science and Technology, Japan, 96, 1984.
20 Asahi Kasei, Jap. Pat. (Examined), 89-8,086; US Pat. 4,415,726; 4,496,505.
21 Griehl W, and Versäumer H, *Faserforsch. u. Textiltech.*, **9**, 226, 1958.
22 Hamana I, in *Formation of Fibers and Development of their Structure: (I) Melt Spinning*, Kagaku-Dojin, Tokyo, 123, 1969.
23 Toyobo, Jap. Pat. (Examined), 65-11,767.
24 Toray, Jap. Pat. (Examined), 87-223,314.
25 Toray, Jap. Pat. (Examined), 70-1,932.
26 Teijin, Jap. Pat. (Examined), 80-10,684.
27 Kikutani T, Morinaga H, Tukaku A and Shimizu J, *Intern. Polym. Processing*, **5**, 20, 1990.
28 Du Pont, Jap. Pat. (Laid open), 68-169,513.
29 Toyobo, Jap. Pat. (Laid open), 89-162,820.
30 Cuculo J A, Tucker P A and Chen G-Y, *J. Appl. Polym. Sci., Appl. Polym. Symp.*, **47**, 223, 1991.
31 Du Pont, Jap. Pat. (Examined), 66-7,892; GB Pat. 1,006,136.
32 Hoechst-Celanese, Jap. Pat. (Examined), 88-528; US Pat. 4,101, 535.
33 Nagai A, Fujiwara T and Murase Y, Preprints of the 10th Meeting of Sen-i Rengo Kenkyu, 96, 1984.

3

Solution spinning

Takashi Tsurumi
Asahi Chemical Industries Ltd, Fuji, Japan

3.1 Introduction

Solution spinning can be classified primarily into two methods, dry spinning and wet spinning. Dry spinning is where polymer solution is solidified through evaporation of polymer solvent.

Wet spinning can be divided further into three methods based on three different physicochemical principles: the liquid-crystal method, the gel method and the phase-separation method. In the liquid-crystal method, a liquid-crystalline solution of a lyotropic polymer is solidified through the formation of a solid crystalline region in the solution. In the gel method, polymer solution is solidified through the formation of intermolecular bonds in the solution. This phenomenon is called gelation. Gelation is caused by a temperature or concentration change in the solution. In the case of phase separation, two different phases appear in the solution, one polymer-rich, the other polymer-lean.

Figure 3.1 shows the phase diagram of a ternary system consisting of polymer, solvent and precipitating agent (non-solvent). The curve represents the boundary between a homogeneous, one-phase system (above the curve), and a heterogeneous, two-phase system (under the curve). In dry spinning, the spinning dope (SD) increases in polymer concentration along the line from SD to pure polymer. In the region above the line A, the spinning dope decreases in polymer concentration and solidification never occurs. To the right of the line B, the polymer concentration increases and the spinning solution solidifies through the formation of a gel (or oriented crystals in the case of lyotropic polymer). In the region between the lines A and B, phase separation occurs, yielding polymer-rich and polymer-lean phases. As is clear from the introduction above, the gel method and the liquid crystal method result in the formation of a solid homogeneous phase as the initial structure of the fiber, but in the case of phase separation, a heterogeneous two-phase

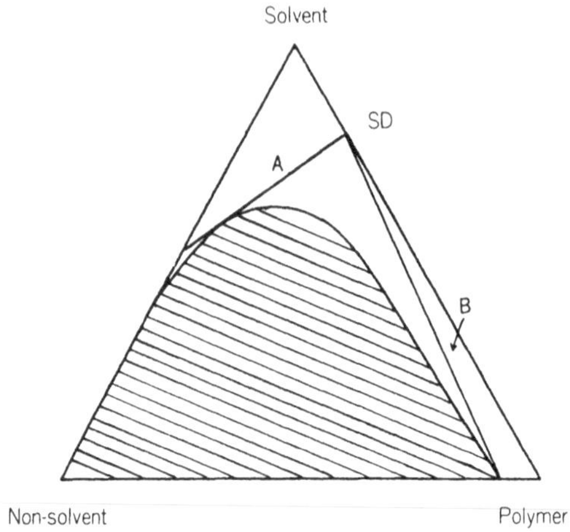

3.1 Phase diagram of a ternary system consisting of polymer, solvent and non-solvent. (SD = spinning dope.)

system appears as the initial structure. This difference in structure significantly affects the choice of fiber-forming method.

In this chapter, the recent development of solution spinning will be presented from both theoretical and applied aspects by reference to the spinning of an acrylic fiber and a Bemberg fiber as examples of wet spinning and of segmented polyurethane fiber as an example of dry spinning.

3.2 Spinning technology of an acrylic filament yarn

3.2.1 Spinning theory for an acrylic fiber

Recently, significant developments have been made in the field of phase equilibrium and phase separation of polymer solutions, which have led to a theoretical understanding of the spinning technology of acrylic fibers.

Figure 3.2 shows the phase diagram of a two component solution of polymer and solvent.[1] If the temperature of a homogeneous polymer solution at point E is lowered to that at point F, the solution becomes thermodynamically unstable, which leads to the generation of 'concentration fluctuation'. The solution becomes visibly cloudy, so the point F is called the 'cloud point'. The curve which links cloud points is called the 'cloud point curve'. A further drop of solution temperature beyond the point F leads to separation of the solution into two phases, a polymer-lean phase (A) and a polymer-rich phase (B).

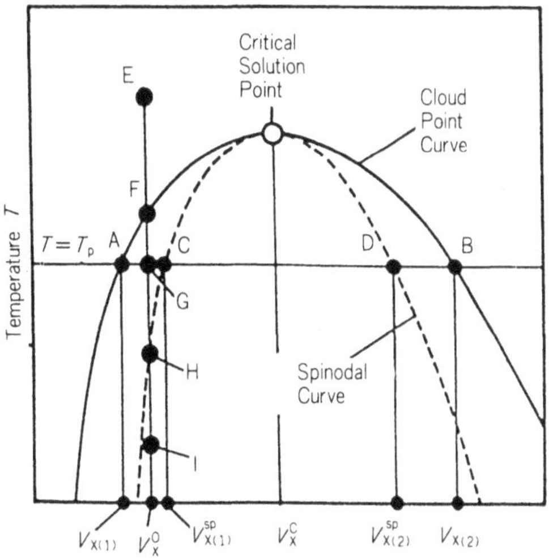

3.2 Phase diagram of a binary system consisting of polymer and solvent.[1]

It has been proved by Kamide and Manabe that this phase separation occurs according to the process called 'micro-phase separation'. Figure 3.3 shows this process diagramatically.[2] In the polymer solution, critical nuclei are formed because of concentration fluctuation if the temperature of the solution is lowered to a temperature below the cloud point. The critical nuclei grow to larger particles called 'primary particles'. If the polymer concentration of the solution (V_x^0) is lower than the concentration at the critical solution point (V_x^c), the polymer-rich phase is a dispersed phase. On the other hand, if V_x^0 is higher than V_x^c, the polymer-lean phase becomes the dispersed phase. Afterwards these primary particles collide and amalgamate with each other and grow to larger particles called 'secondary particles', which further amalgamate and fuse

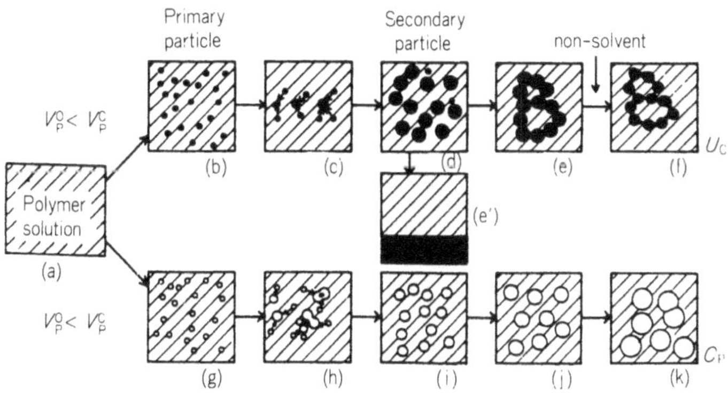

3.3 Diagram of micro-phase separation.[2]

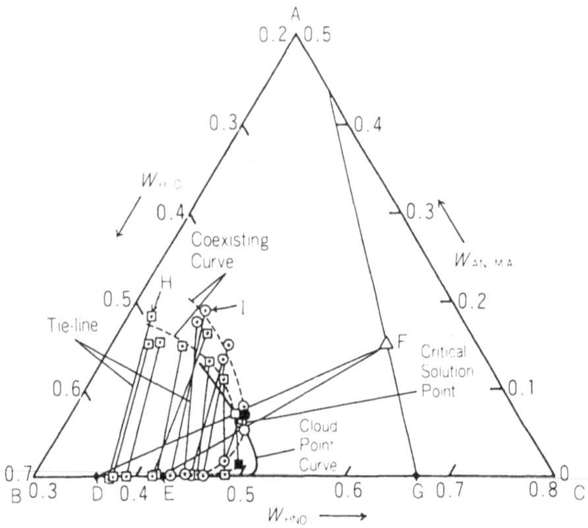

3.4 Phase diagram of a ternary system consisting of polyacrylonitrile, nitric acid and water.[3]

to create the fiber structure. Such micro-phase separation phenomena can be seen only in the case of a three component solution of polymer, solvent and nonsolvent. Figure 3.4 shows a phase diagram of such a three component system comprising poly(acrylonitrile/methyl acrylate) copolymer, solvent (strong aqueous nitric acid solution) and nonsolvent (water).[3] If a dilute aqueous nitric acid solution (coagulant) E is added to a spinning solution F, the composition of a mixture changes along the line connecting F and E, crossing the cloud point curve at the point marked by a white circle. As the polymer concentration at this point is lower than that at the critical solution point, marked by a dotted white circle, the polymer-rich phase forms a dispersed phase. On the other hand, if a more dilute aqueous nitric acid solution (coagulant) D is added to the solution F, the concentration of the resulting mixture changes along the line connecting F and D, crossing the cloud point curve at the point marked by an open square. As the polymer concentration at this point is higher than that of the critical solution point, in this case the polymer-lean phase forms a dispersed phase and the polymer-rich phase forms a continuous phase.

Figure 3.5 shows the relationship between the concentration of coagulant and the maximum draft ratio (dr_{max}), that is, the ratio of the fiber velocity after the coagulation bath to the velocity of the spinning dope at extrusion from a spinneret. When the concentration of coagulant increases, at first dr_{max} decreases gradually and then it increases rapidly after reaching a minimum. This minimum concentration is called the critical concentration. Generally, the method of spinning with coagulant

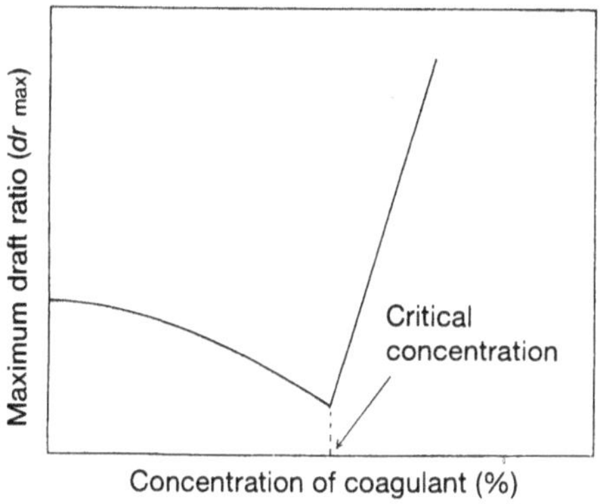

3.5 Dependence of maximum draft ratio on concentration of coagulant.

concentration below this point is called 'low concentration spinning' while that with coagulant concentration above this point is called 'high concentration spinning'. This concentration is considered to correspond to the 'critical solution point' in Fig. 3.4. When spinning is carried out in the low concentration region, a continuous layer (skin) appears on the surface of the spun dope, then the nonsolvent in the coagulant diffuses through this skin into the spun dope inducing coagulation inside it and a change in volume. As the skin is rather rigid, it cannot deform in proportion to the shrinkage of the inside volume caused by the amalgamation of polymer particles. This is the reason macro-voids are formed inside the fiber. The more the concentration of the coagulant decreases, the greater the number and the larger the voids that are formed inside the fiber. These voids often induce opacity in a fiber. If the elongation and drying of a fiber are properly carried out, a transparent fiber without voids can be obtained.

On the other hand, if spinning is carried out in the high concentration region, the coagulated fiber does not have a skin but has a structure formed uniformly by aggregation of polymer particles. In this case, coagulation does not proceed as rapidly as in the case of low concentration spinning, but because of the lack of skin, both solvent and nonsolvent diffuse smoothly between the inside of the fiber and the external coagulant, which makes the fiber structure uniform. In addition, as the phase separation occurs in the coagulant at a higher concentration of solvent than in the case of low concentration spinning, the polymer particles contain a larger amount of solvent and are more easily elongated and fused with each other under extensional forces in the spinning process than in low concentration spinning. However, high concentration spinning has some defects. Firstly, a fiber of this type has very low tenacity in the coagulation bath and is easily broken by extensional forces. Secondly, a larger amount of polymer comes out into the coagulant than in low concentration spinning.

In the history of the development of acrylic fibers, the manufacturing method for staple fiber was developed first, then that for filament yarn on the basis of the manufacturing technology for staple fiber. The manufacturing technology for staple fiber is that of the manufacture of fibers in the form of a tow which has a total denier of several tens of thousands, so the spinning velocity is generally less than a few hundred meters per minute. On the other hand, much effort has gone into development of high-speed spinning of continuous-filament yarn, which will now be discussed.

3.2.2 Spinning technology of an acrylic filament yarn

There are two methods of spinning an acrylic fiber (Fig. 3.6). One is an immersed-jet method where spinning dope is extruded through a spinneret which is immersed in a coagulation bath. The other is a dry-jet wet spinning method (or air-gap method) where a dope is extruded through a gaseous atmosphere from a spinneret which is set above a coagulant bath. In the case of the immersion method, the spinning dope starts coagulating immediately after it is extruded from a spinneret, so it is difficult to draft filaments in the coagulation bath at a high rate, especially in the case of low concentration spinning. On the contrary, it is easy to draft extruded dope at a high rate while it is in the air in the case of the air-gap method. In addition, filaments made by this process have a smooth and glossy surface compared with those made by the immersion process. The manufacturing technology for acrylic filament yarn has been developed on the basis of these two spinning technologies.

3.2.2.1 Immersion method

Figure 3.7 shows a spinning apparatus utilizing the immersion method.[4] A spinning dope made of polyacrylonitrile or an acrylonitrile copolymer (hereafter referred to as PAN) dissolved in *ca* 70% nitric acid solution is

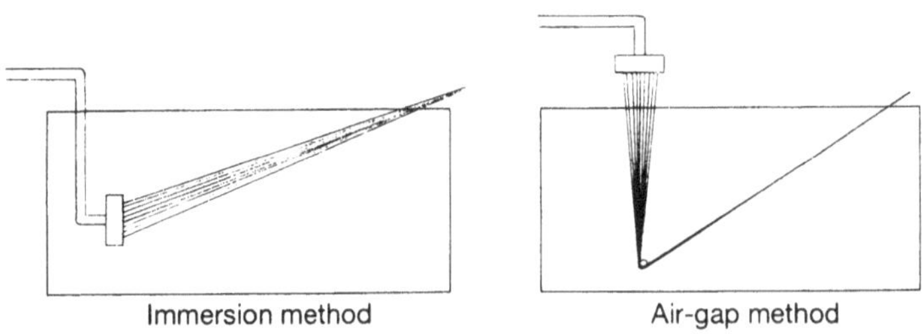

3.6 Spinning methods for acrylic fiber.

extruded into a coagulant bath containing *ca* 30% aqueous nitric acid solution at −3°C. After coagulation, the fiber is drawn out from the coagulant bath by means of a set of rollers (14) rotating at 5–10 m/min. After complete removal of solvent in the washing bath (6), the fiber is passed by means of a set of rollers (15) to an elongation bath (7) which is filled with hot liquid or steam. The fiber is elongated at least four-fold between rollers (15) and (16). The fiber thus oriented is then dried, first in a dryer (9) after treatment with oil or size, then in a second dryer after being treated with finishing agent, and finally wound up on a winder (13). The reason for drying the fiber in two stages is to obtain good mechanical properties. In the first dryer, the fiber shrinks by 5–10%; in the second dryer it is elongated by 0–10%. By controlling the shrinkage and elongation rate, good mechanical properties can be obtained, as can be seen in Table 3.1.

The characteristics of the fiber obtained by this process are an excellent lustre and a handle like that of silk. This silky handle is derived from micro-stripes on the fiber surface. This fiber has high tensile strength and low tensile elongation.

3.2.2.2 Air-gap spinning method

Figure 3.8 shows a spinning apparatus utilizing the air-gap spinning method developed by the Monsanto company.[5] The following is an example of spinning conditions in such a process.

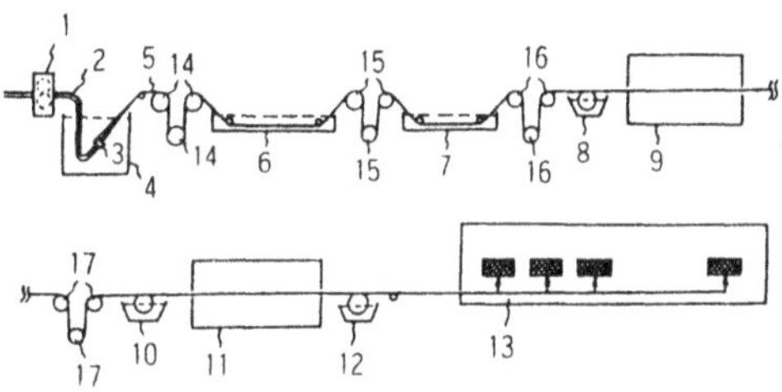

3.7 Spinning apparatus using immersion method.[4]

Table 3.1 Effect of shrinkage and elongation in dryers

Shrinkage in the first dryer (%)	0	5	10	15	19	19
Elongation in the second dryer (%)	0	0	1	3	0	4
Tensile strength (g/d)	3.9	3.8	3.6	3.6	3.1	3.4
Tensile elongation (%)	8.0	10.7	15.0	16.8	28.0	21.3
Knot strength (g/d)	1.1	1.9	2.1	2.2	3.1	2.5
Knot elongation (%)	1.8	6.0	7.0	10.0	22.4	15.0
Shrinkage at the boil (%)	14	12	11	10	4	6

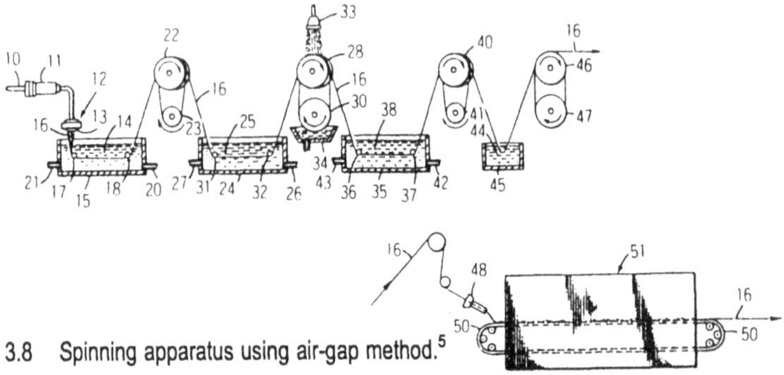

3.8 Spinning apparatus using air-gap method.[5]

Spinning dope made from a 26% PAN solution in N,N′-dimethylacetamide (DMAc) is extruded from the spinning nozzle (13) into the air. After running for 1.3 cm in the air, it enters a coagulant bath containing 50% aqueous DMAc solution. The fiber is elongated in the coagulant bath to some extent by the tension applied by a roller (22), then removed from the coagulation bath at a velocity of 9.3 m/min. Then it is elongated 6.1 times in an elongation bath containing hot water at 100°C. The fiber is cleared of solvent with hot water at 50–80°C while it runs 30–40 times over rollers (28 and 30). The fiber is dried on drum dryers (46 and 47) after being treated with oil in a bath (45). The fiber velocity on the drums is about 46 m/min.

The characteristics of this process are air-gap spinning and drying without tension after shrinking the elongated fiber in hot water. By this process, a fiber with a smooth glossy surface and high toughness can be obtained.

3.2.3 High-speed spinning of an acrylic filament yarn

3.2.3.1 Technological requirements to achieve high-speed spinning

The following requirements should be fulfilled in order to achieve high-speed spinning in the manufacture of an acrylic filament yarn:

1 Rapid coagulation. In acrylic fiber manufacture the fiber is elongated by up to several tens of times, after coagulation, to obtain appropriate mechanical properties through molecular orientation and densification. The fiber velocity in the coagulation bath is not so high as that for a regenerated cellulose, which is not substantially elongated after coagulation if the same winding velocity is used, but much higher speed is necessary compared with a conventional acrylic spinning method.
  The following are measures required to achieve high speed coagulation:
  (a) In the case of an inmersion method:
    (i) High velocity of spinning dope at extrusion when spinning in

a low-concentration spinning region.
   (ii) High velocity of spinning dope at extrusion and a large draft ratio when spinning in a high-concentration spinning region.
 (b) In an air-gap spinning method, high velocity of spinning dope at extrusion and a large draft ratio.

   Resulting from these measures, the maximum velocity of a fiber in a coagulation bath is only a few tens of m/min in the case of low-concentration spinning utilizing an immersion method due to the increase of pressure behind the spinneret, an increase of the fiber tension in the coagulation bath, and low dr_{max}. Consequently, the other two methods have been employed in the development of high-speed spinning.

2 Rapid removal of the solvent. As mentioned before, elongation of the fiber after coagulation is indispensible in acrylic fiber spinning in contrast to regenerated cellulose-fiber spinning. This permits relatively low fiber speeds in the coagulation bath for acrylic fiber spinning but makes it difficult to utilize a net conveyer system after removing the solvent.

   The following are the measures for rapid removal of solvent using high-speed spinning:

   (a) Low resistance of washing liquid and high efficiency of washing in a washing process such as the Hoffman type (straight type), or
   (b) sufficient retention time of fiber on Nelson-type rollers.

   The former method is advantageous from the viewpoint of saving space and cost of equipment, but it is difficult to balance the washing effect with the resistance of the washing liquid. The latter method is advantageous from the viewpoint of balance but requires larger and more expensive equipment.

From the viewpoint of mechanical properties it is desirable to dry the fiber at low tension. The following measures decrease fiber tension while drying:

1 Low applied tension while drying on a drum-type dryer and during any subsequent relaxation of the fiber to reduce residual tension.
2 Drying the fiber on Nelson rollers, if necessary with tapering rollers.
3 Drying the fiber without tension on a net conveyor.

3.2.3.2 High-speed spinning in the high concentration region with an immersion-spinning method

Figure 3.9 shows a diagram of a high-speed spinning apparatus for spinning in the high concentration region using an immersion-spinning method of the Monsanto Company.[6] The following is an example of spinning conditions used in this process.

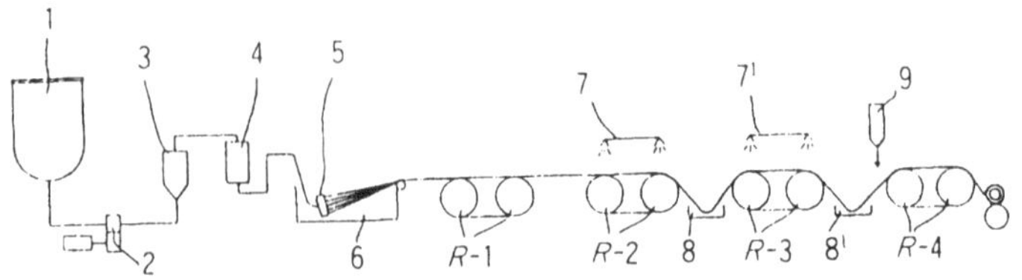

3.9 High-speed spinning apparatus: immersion method in high-concentration region.[6]

A spinning dope made from a 25% PAN solution in DMAc is extruded into a coagulant bath containing 83% aqueous DMAc soution at 30°C and is drawn out from the bath by means of the rollers R-1 and R-2 in an air atmosphere. The solvent is removed from the fiber on the rollers R-2 and R-3 with deionized water at 70°C. The fiber is elongated with steam at 105°C (8) between washing rollers. The fiber is then elongated again (8′), which is heated electrically at 250°C, treated with a finishing agent, dried on the rollers R-4, and wound up on a bobbin.

In this process the tension of the running fiber is controlled at an appropriate level, especially during washing and drying when it is kept at a low level. If the fiber is elongated in the tube 8′, a fiber with high strength and low elongation is obtained, but if the fiber is shrunk in this tube a fiber with low strength, high elongation and low shrinkage in boiling water is obtained. Additionally, the shrinkage in boiling water is lowered if the fiber on the bobbin is treated with steam. Spinning speeds up to 1000 m/min are available using this process.

3.2.3.3 High-speed spinning with an air-gap method

Figure 3.10 shows a flow sheet for high-speed spinning with an air-gap spinning method of the Mitsubishi Rayon Company.[7] The spinning dope, a PAN solution in DMAc, is extruded downwards from the spinneret, falls 1–20 mm in air and enters a coagulant bath containing aqueous DMAc solution (for example, 70% at 40°C). After coagulation the fiber is washed with hot water, elongated 1.8–4.5 times in boiling water, and then elongated 6–12 times with steam at 2–4 kg/cm^2. After the elongation, the fiber is dried and then shrunk using a hot roller, a hot plate or high pressure steam.

The first elongation, with boiling water, has a great influence on the second elongation, with high pressure steam. Unless the fiber is elongated to the appropriate extent in the first stage, a high elongation rate cannot be obtained in the second stage. If the fiber is elongated 2–4 times in the first stage, a high rate of elongation is possible in the second stage and a maximum velocity of 2160 m/min can be obtained (Table 3.2).

Figure 3.11 shows an example of a spinning apparatus for the air-gap method from Asahi Chemical Industry.[8] By employing an air-gap

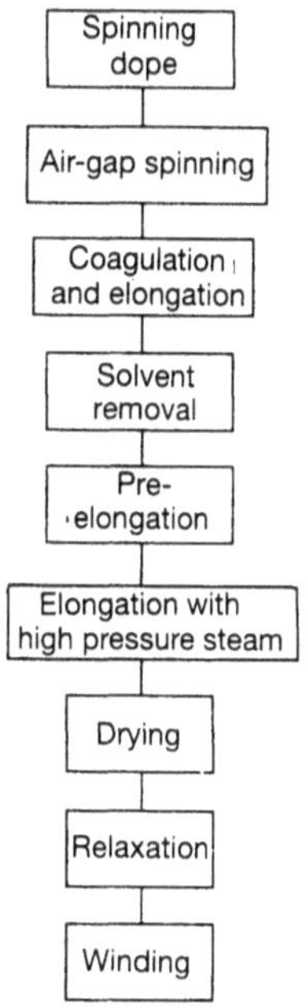

3.10 Flow chart of high-speed spinning: air-gap method.

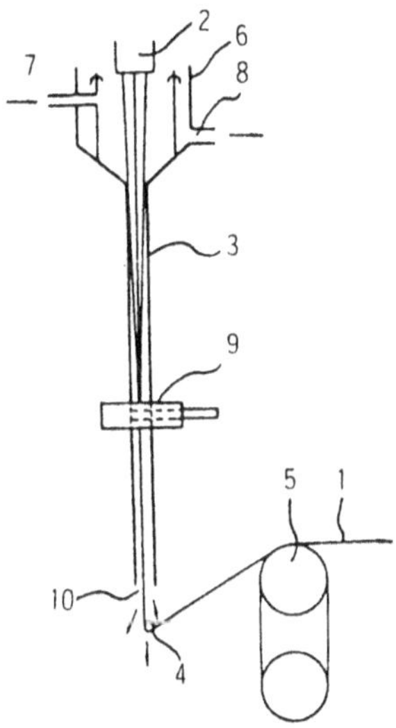

3.11 Spinning funnel for the air-gap method.[8]

Table 3.2 Relationship between spinning speed and fiber properties

First stage elongation ratio	Conditions of pressurized steam elongation			Final spinning speed (m/min)	Fiber properties		
	Steam pressure (kg/cm^2-G)	Maximum elongation speed (m/min)	Stable elongation speed (m/min)		Denier (d)	Tensile strength (*a*d)	Tensile elongation (%)
1.5	2.7	1450	900	810	110	4.10	20.0
2.0	2.9	2430	1700	1530	59	6.31	17.2
3.0	2.0	2400	1600	1440	63	5.90	16.3
3.0	3.2	3020	2400	2160	42	6.82	15.4
3.0	4.0	2370	1500	1350	67	5.50	16.2
4.0	3.5	2260	1500	1350	67	5.43	17.7
5.0	3.6	1330	800	720	125	3.94	18.7

method, the velocity of the spinning dope before coagulation can be increased, but the high resistance of the coagulant liquid has to be overcome. If the resistance is too high, the fiber breaks in the coagulant bath, so high speed cannot be obtained. In this apparatus the coagulant liquid, introduced through an inlet (7), flows down through a spinning funnel (3). In this funnel, excess coagulant liquid overflows into an outer funnel (6) and flows out through an outlet (8), maintaining the liquid head at a desired level. The velocity of the coagulant liquid in the straight part of the funnel can be controlled by changing the inner diameter of the funnel. This downward flow of the coagulant liquid decreases its resistance to the fiber compared with a static coagulant bath, so the fiber is elongated and coagulated smoothly.

By employing this spinning funnel and a net (mesh) conveyer for drying, a spinning velocity of 1000–2000 m/min can be obtained. Table 3.3 shows the mechanical properties of a conventional acrylic filament manufactured by an immersion process in the low-concentration region and that of a filament manufactured by the high-speed spinning process with an air-gap method.[9] The fiber obtained from the high-speed process has higher tensile elongation, knot strength and abrasion strength and lower shrinkage in boiling water than the conventional fiber. These advantages of the fiber from the high-speed process are the result of employing an air-gap method and drying without tension on the net conveyor. Figure 3.12 shows an SEM of the surface of these fibers. The conventional fiber, (a), has micro-stripes on its surface but the fiber made by the high-speed spinning method, (b), has a smooth surface without micro-stripes. The micro-stripes are thought to be formed through depression of voids in the fiber which are common in fibers made in the low-concentration region when the fiber is elongated in the coagulant bath. On the other hand, the fiber made in the high-speed spinning process is elongated mainly in air just after extrusion from the spinneret and only slightly in the coagulant bath. This is thought to be the reason why this fiber has a smooth surface.

Table 3.3 Properties of fiber from the conventional spinning method and the high-speed spinning method

	Conventional	High-speed spinning
Tensile strength (g/d)	5.1	3.0
Tensile elongation (%)	13.9	27.8
Knot strength (g/d)	2.6	2.0
Knot elongation (%)	7.5	25.2
Maximum twist number (t/m)	1092	1928
Shrinkage in boiling water (%)	6.8	1.5
Fibrillation grade*	2nd	5th

* Fibrillation grade: evaluation was done by means of a 5-grade scale. The 5th grade is the highest.

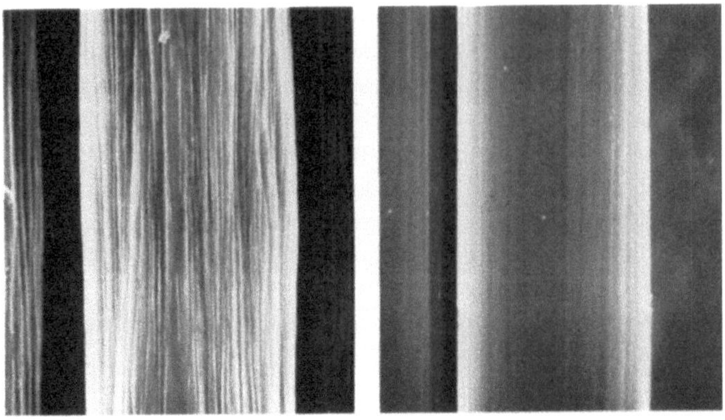

(a) (b)

3.12 Scanning electron micrograph of surfaces of acrylic fibers: (a) conventional fiber; (b) fiber made by high-speed spinning process.

Acrylic filament yarn has been used rarely in the high fashion field. However, as the high-speed process has realized large improvements both in quality and cost, it will in future be widely used in general textile and industrial fields, like regenerated cellulose filament yarn which is already used in such fields.

3.3 Spinning technology of Bemberg rayon

3.3.1 Development of manufacturing technology of Bemberg rayon

It was shown in the middle of the nineteenth century that cellulose could be dissolved in ammoniacal solutions of copper hydroxide. The industrial technology for manufacturing Bemberg rayon (ISO term cupro, FTC term cupra, i.e. cuprammonium-regenerated cellulose; hereafter referred to as Bemberg rayon) was developed in 1918 by J P Bemberg in Germany. Thereafter, many companies worldwide introduced this technology and started the production of Bemberg rayon. Nowadays only a few companies, including Asahi Chemical Industry in Japan and Bemberg SpA in Italy, continue to produce Bemberg rayon. This is due to its lack of competitiveness against viscose rayon arising from the use of expensive materials such as copper and ammonia to dissolve the cellulose.

In manufacturing Bemberg rayon, cotton linters are used as the raw material. Cotton linters are short fibers which are rubbed off cotton seeds after cotton fiber is cut from them. Refined linters are obtained by boiling raw cotton linters with alkali solution and then bleaching them. The refined linters are characterized by a narrow distribution of degree of polymerization, chemical purity and low content of oxidized groups.

Copper hydroxide is dissolved by aqueous ammonia, forming a complex salt (tetra-ammonium copper hydroxide). The refined linters are added to copper ammonium solution which contains copper hydroxide as a precipitate. Cellulose forms a complex with tetra-ammonium copper hydroxide which dissolves in the solution. Meyer[10] proposed the following chemical formula for the complexed product.

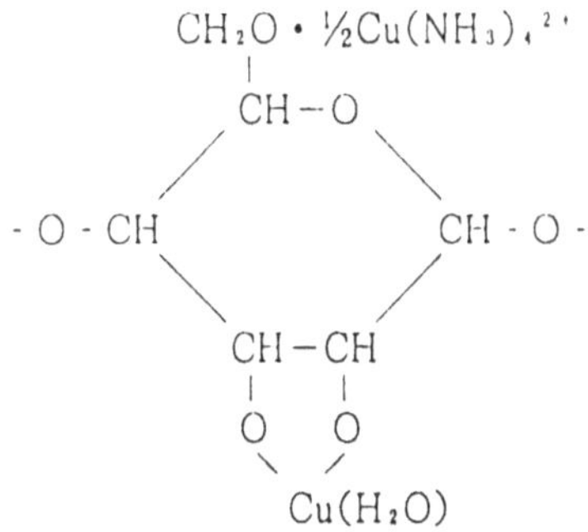

The spinning technology of Bemberg rayon is based on the method developed by Thiele in 1901 in Germany. The spinning dope extruded from a spinneret flows down with hot water (coagulant) through a spinning funnel in the shape of a cone (Fig. 3.13).[11,12] In the funnel the

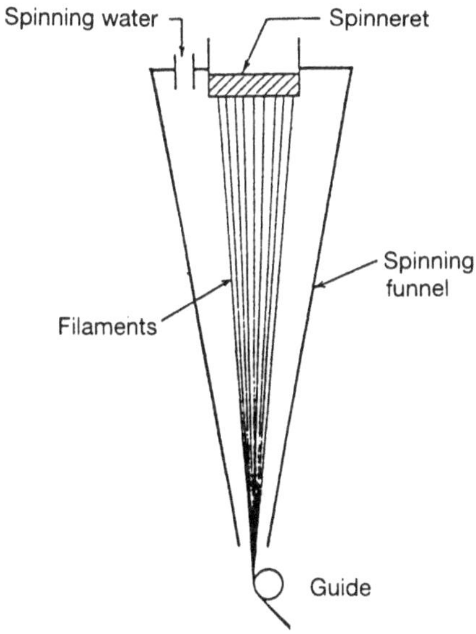

3.13 Stretch-spinning method.[12]

spinning dope is coagulated gradually while being elongated several hundred-fold and thus the fiber is formed.

3.3.2 Review of spinning methods

3.3.2.1 Hank-spinning method

The hank-spinning method is one of the oldest Bemberg spinning methods and is a batch process which uses a winding frame called a Hank in Gemany (Fig. 3.14). The fiber on the Hank is taken off it in the shape of a ring and then regenerated and dried. The spinning speed is 40–80 m/min.

The cellulose cuprammonium solution is extruded from a spinneret with a diameter of 0.6–1.0 mm into hot water (the spinning water) in a spinning funnel. The extruded spinning dope flows down in the funnel, being coagulated and elongated at the same time. Generally, the degree of coagulation depends on the ratio of the ammonia and copper which is removed from the fiber to the amount of ammonia and copper initially present in the fiber.

Figure 3.15 shows the changes in the velocity of the spinning water and the fiber in the funnel. In the upper part of the funnel, spinning dope extruded from the spinneret flows down under gravity, gradually growing thinner by elongation. During this time, ammonia diffuses into the hot water through the surface of the fiber and so coagulation starts from the surface. The elongation at this stage causes cracking of the surface layer, diffusion of ammonia from the new surface layer ensues and coagulation proceeds. Through repetition of these processes, coagulation and elongation continue. When the coagulation rate reaches a certain point,

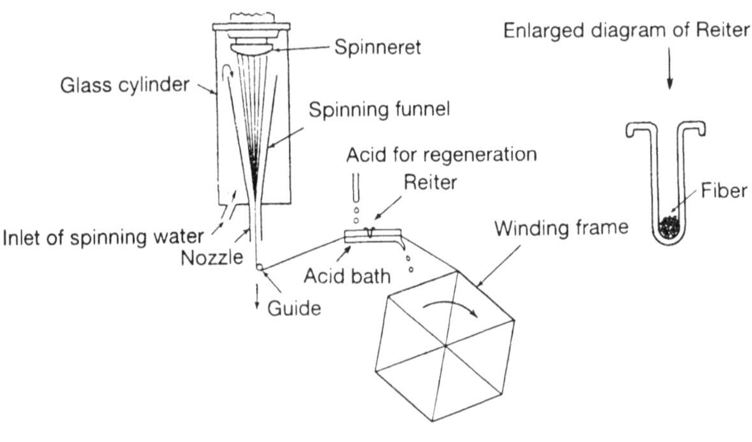

3.14 Hank-spinning method.

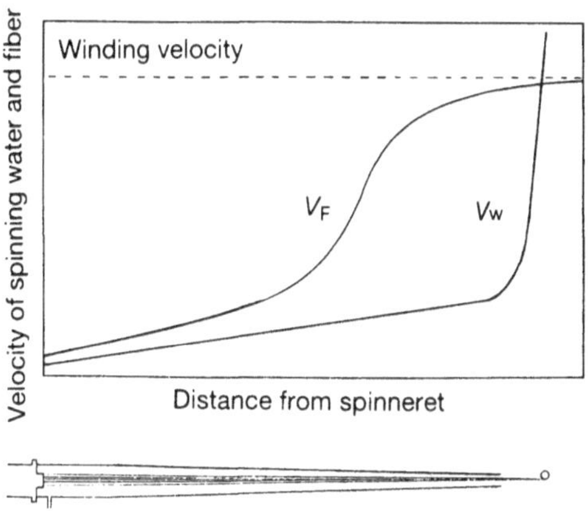

3.15 Change of velocity of spinning water and fiber: V_w = velocity of spinning water; V_F = velocity of fiber.

the fiber changes to an insoluble complex called a blue yarn. The viscosity of the fiber rises rapidly and it loses its fluidity. In the middle section of the funnel, cellulose molecules, which still form a complex with copper, are oriented longitudinally by means of a winding force transmitted from the lower part of the funnel. In the lower part of the funnel, the fiber is elongated only slightly by the winding force. After the fiber leaves the funnel outlet, it is regenerated with acid and dried in later processes without substantial elongation. The fundamental structure of the fiber, including the orientation of molecules and the degree of crystallization which contol the fiber properties, is considered to be determined at the stage of the formation of the blue yarn. So the extent of elongation in the upper part of the funnel and the tension imposed on the blue yarn from its formation to the stage of regeneration have a decisive influence on the fiber properties.

Bozza and Elsasser thoroughly researched this spinning process. Bozza[13] expressed the differential increase of fiber velocity dv_f by the following equation using fiber velocity, v_f ,at a distance x from a spinneret, tension imposed on the fiber at that point f, cross section area of the fiber q and viscosity μ

$$dv_f = (1/3\mu)\,(f/q)\,dx$$

Elsasser[14] determined the tension f and viscosity μ of the fiber at a series of distances from the spinneret by inserting dv_f/dx into this equation calculated from the diameter of the fiber in the funnel, which was measured directly. From these results he introduced the idea of 'optimum coagulation state' and concluded that the tension imposed on the fiber at the optimum coagulation state determines the strength of the fiber.

According to his calculation, the fiber at this state has a viscosity of 56 500 poise which equals that of rather soft asphalt. Therefore, it is considered that the fundamental structure of the fiber is formed at an early stage of coagulation in this spinning method.

The blue yarn which comes out from the bottom of the funnel with the spinning water changes its direction and is separated from the water by means of a guide set under the funnel, and is wound up on a frame after being treated with 6% aqueous sulfuric acid solution for final regeneration. This solution is poured on to the fiber on the frame during winding to remove ammonia and copper.

The Hank yarn is manufactured without twist and is woven or knitted either without twist or after being twisted. There is a unique feature of the Hank process called Reiter colligation that makes it possible to handle Hank yarn without twist in the spinning process and the later processes. The Reiter is a small instrument set just before the apparatus for sulfuric acid treatment (Fig. 3.14). When the blue yarn comes into contact with the sulfuric acid solution, rapid regeneration occurs with generation of active OH groups and elimination of water from the yarn. If filaments in the yarn are in close contact with each other, they bond to each other by means of chemical bonds based on OH groups. The Reiter with an appropriate curvature at its bottom for the passage of the yarn is set at the point where regeneration of the yarn takes place most intensively in order to get the filaments into close contact with each other and bind them with OH bonds. These OH bonds are formed intentionally so that they ensure that yarns can be handled easily in the manufacturing processes during spinning, knitting or weaving. At the same time they also ensure that filaments can move freely enough apart in the fabric to give it soft touch and good uniform appearance. For this reason, the Reiter colligation should be reversible. The extent of colligation is controlled by changing the position and the curvature of the Reiter.

The yarn wound on the Hank is then conveyed to a regeneration apparatus where it is further regenerated, washed thoroughly for several hours, treated with finishing oil and then dried in a dryer for several hours.

In the Hank process, the spinning speed is restricted by two factors. One is the use of a Hank frame for winding. If the spinning speed is increased, the increased tension of the yarn tightens the yarn wound on to the frame and binds the filaments together both within the yarn and between the yarns, which leads to insufficient regeneration and drying. The other restrictive factor is the spinning condition in the funnel. These resrictions are overcome by the following continuous spinning method.

3.3.2.2 Continuous-spinning method

Two types of continuous-spinning apparatus, the Hoffman type and the Duretta type, are known thus far. The main differences between them are in the regeneration and drying processes.Only the Hoffman process will be described here.

In the Hoffman apparatus, the fiber, which comes out of the funnel, runs straight through the regeneration stage and the drying apparatus and is wound up continuously on a winder. The spinning speed in this process is 100–150 m/min.

One of the technological improvements which make a higher speed possible is the employment of the double spinning-funnel method (Fig. 3.16).[15] In the single spinning-funnel method of the Hank process, an increase of spinning speed requires:

1 Improvement of funnel dimensions.
2 Changes in the amount and temperature of the spinning water.

However, if these conditions are changed to facilitate elongation at high speed, the fiber comes out of the funnel uncoagulated.

This problem is solved in the double spinning-funnel method by separating the role of spinning water into coagulation and elongation. In this method, the temperature of the 'first' spinning water used for the upper funnel is lower than that in the Hank method, to facilitate elongation, but the temperature of the second spinning water used for the lower funnel is higher to ensure sufficient coagulation. Additionally, in the lower funnel, turbulent flow is generated by mixing the first water and the second water; this accelerates coagulation.

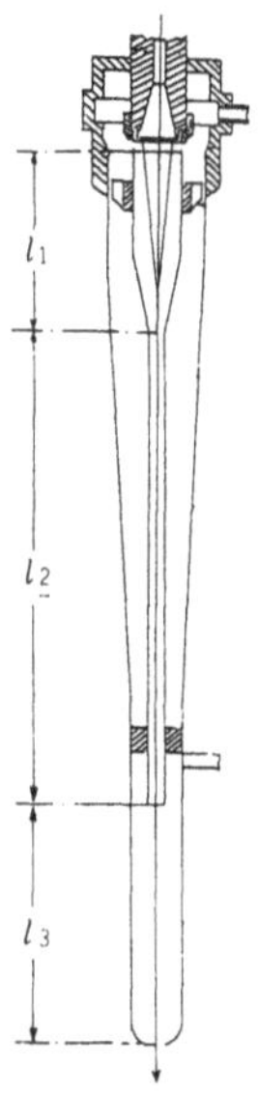

3.16 Double spinning funnel.

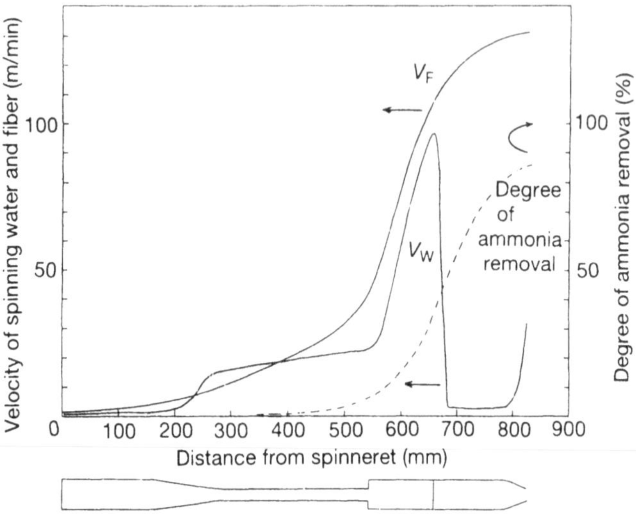

3.17 Change of velocity of spinning water and fiber: double spinning funnel.

Figure 3.17 shows the change in velocity of the spinning water and the fiber and also the degree of coagulation in the funnel as a function of the distance from the spinneret. The degree of coagulation is very low up to the outlet of the upper funnel but rises rapidly after the fiber reaches the second water. If the temperature of the first water is too high, coagulation proceeds too far and copper hydroxide from the fiber sticks to the outlet of the upper funnel. This causes problems such as a change in velocity of the water and breakage of the fiber. On the other hand, if the temperature is too low, insufficient coagulation causes problems such as breakage of the fiber and deterioration of fiber properties.

The other technological improvement required for a continuous-spinning process at high speed is an acceleration of regeneration. In the Hank-spinning method, it takes many hours to regenerate and dry fibers as they are tied up closely in a bundle. On the other hand, in the Hoffman process, the regeneration is finished in a few seconds because the fiber runs straight through a regeneration bath which, in any case, should not be too long because of the cost of the machine. In order to accelerate regeneration, replacement of water around the fiber is important. It is also essential to keep the fiber structure in a state which facilitates diffusion of water and of ions such as sulfuric acid and copper. Figure 3.18 shows a diagram of part of the regeneration apparatus. The sulfuric acid solution flows in the regeneration bath against the flow of the fiber. The water layer which covers the yarn is removed by suppressing and supporting dams. A most important point is not to keep the filaments in the yarn bound together but to keep them apart from each other in the

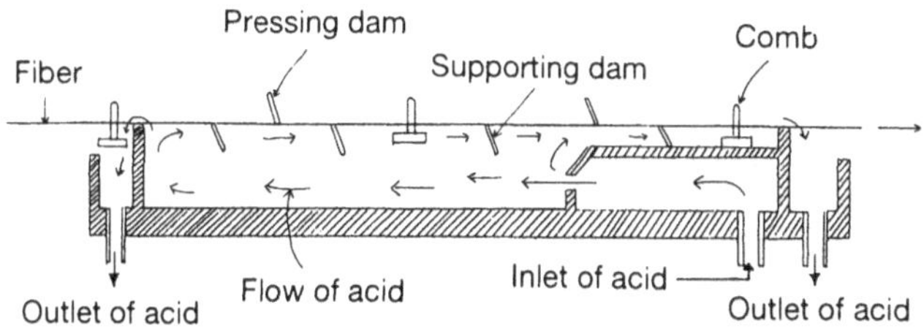

3.18 Regeneration apparatus of Hoffman continuous-spinning method.

bath. The water layer can then be taken off smoothly and replaced with fresh water which has lower ammonium and copper ion concentration. This condition can be attained by ensuring that the coagulation level is sufficiently high before the fiber reaches the regeneration bath.

In addition to these factors, it is important to keep the fiber structure, especially that of the surface, rather porous in order to accelerate diffusion of ions in the regeneration bath. Figures 3.19, 3.20 and 3.21[16] show the dependence of the degree of gloss, swelling and dye absorbancy on the concentration of sulfuric acid which the yarn encounters first in the process. It is clear from these Figures that a high concentration of sulfuric acid makes the fiber dense. Figure 3.22 shows the dependence of residual copper concentration in the fiber on the concentration of sulfuric acid in the first acid bath. As the acid concentration increases, the

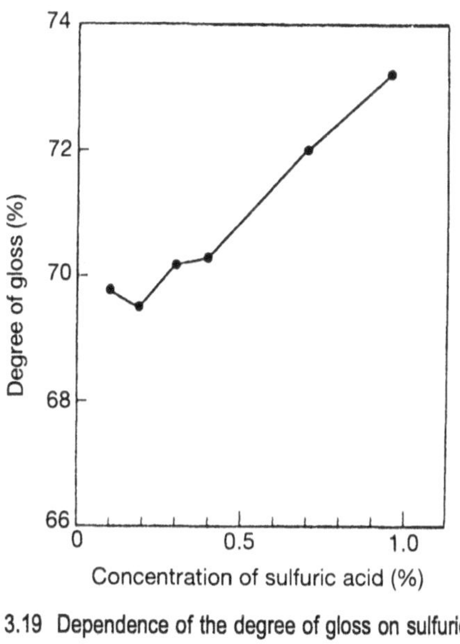

3.19 Dependence of the degree of gloss on sulfuric acid concentration.[16]

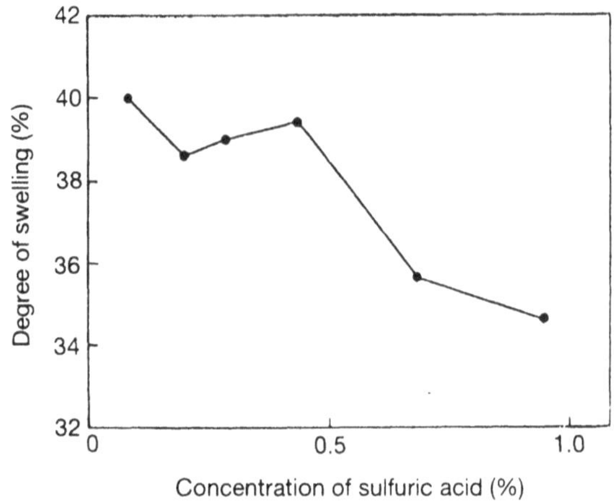

3.20 Dependence of the degree of swelling on sulfuric acid concentration.[16]

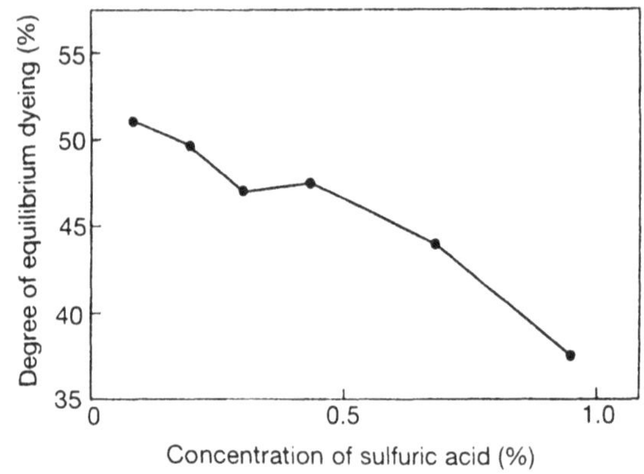

3.21 Dependence of the degree of equilibrium dyeing on sulfuric acid concentration.[16]

copper concentration of the fiber just after the first acid bath decreases but that of the final product increases. This means that the acid concentration of the first acid bath should be kept low, for example lower than 0.5%, in order to attain a low level of copper concentration in the final product.

The Hoffman-type spinning method is superior to the Hank-spinning method because it is continuous and permits higher spinning speed. However, it still has the following drawbacks:

1 Low spinning speed: almost twice as high as that of the Hank process, but much lower than that of synthetic fibers. Higher speed brings about deterioration of fiber properties and leads to fiber breakage in the process.

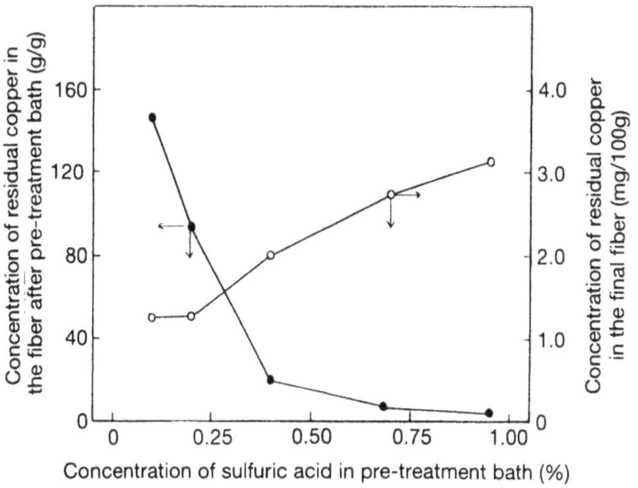

3.22 Dependence of the residual concentration of copper in the fiber on sulfuric acid concentration.

2 Frequent contact of the fiber with parts of the apparatus causes filament breaks (generation of fluffs) and even yarn breaks.
3 Problems with residual copper: even slight changes of spinning condition tend to cause increased residual copper which has a large influence on the dyeability of the fiber.
4 Due to the regeneration and drying under high tension, the tensile elongation is small and shrinkage at the boil is large. These properties are sometimes useful during handling but restrict the fiber's range of use.

In order to overcome the drawbacks of this process and realize a much higher spinning speed, the following technological requirements should be fulfilled:

1 Coagulation and elongation at much higher speed.
2 Regeneration and drying under no tension or low tension.
3 Continuous process from spinning to winding.

These requirements are satisfied by the following high-speed spinning method.

3.3.3 NP-type spinning method

3.3.3.1 High-speed spinning technology in the NP method

A spinning speed of about 400 m/min is realized in the NP method. In the double funnel method of the Hoffman process, the following conditions lead to the highest spinning speeds:

1 Lower temperature and larger amount of the first spinning water.
2 Higher temperature and larger amount of the second spinning water.

However, as the spinning speed rises, the resistance of the spinning water to the fiber in the second funnel increases, the fiber properties deteriorate and many fluffs are generated. The limit of spinning speed with the Hoffman double funnel is about 200 m/min.

Figure 3.23 shows the NP-type spinning apparatus. This apparatus comprises an upper funnel, an instrument called a CJ (short for 'coagulation jet') and a lower funnel. The upper part of the upper funnel contains a mesh for smoothing the flow of the spinning water in order not to entangle filaments just extruded from the spinneret and not to enhance coagulation. The shape of the upper funnel is designed so that the filaments can be highly elongated smoothly without enhancing coagulation.

Figure 3.24 shows the inner structure of the CJ. The apertures, with diameters of several mm, open in a circle, the center of which coincides with the center of the running fiber on the undersurface of the CJ. The second spinning water is supplied from these apertures. No further spinning water is supplied to the lower funnel, but the first and second waters flow into it and mix there. A controlled head of water is maintained in this funnel. Figure 3.25 shows the change in velocity of the spinning water and the fiber and the degree of coagulation.

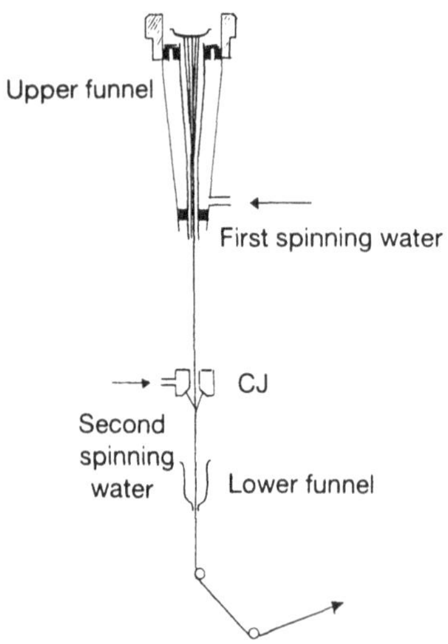

3.23 NP-type spinning funnel.

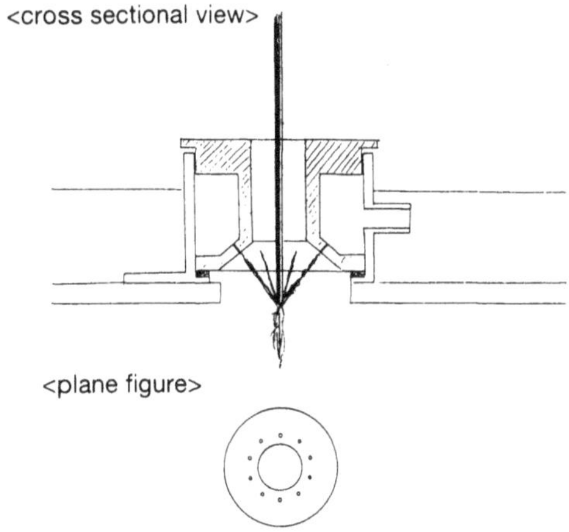

3.24 Structure of CJ water-supply device.

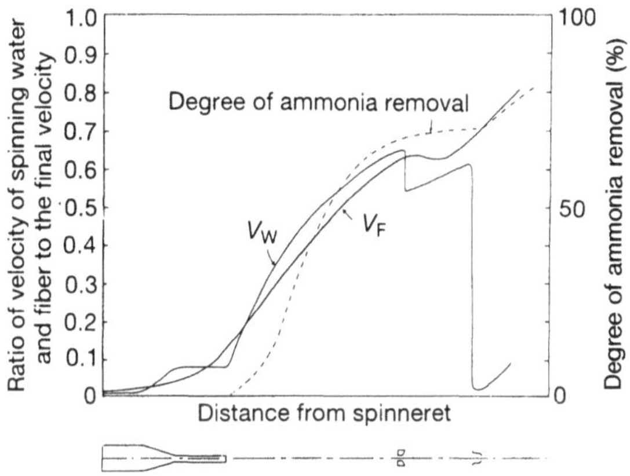

3.25 Change in the velocity of spinning water and fiber as a function of distance from spinneret.

3.3.3.2 Regeneration, drying and winding technology in the NP method

It is not realistic to apply the regeneration and drying apparatus of the Hoffman process to high-speed spinning at several hundreds to 1000 m/min because of the high tension which will produce frequent fiber breaks and high boil shrinkage, and also increase investment cost. In the NP process a net (mesh) conveyer is employed to regenerate and dry fiber (Fig. 3.26).[17]

The apparatus shown in Fig. 3.27 is used for accumulating the fiber which comes out of the funnel at a high velocity. After being treated with sulfuric acid the fiber is drawn by a DKR (double kick roller) down on to a reversing roller. The fiber is oscillated transversely to its direction by an

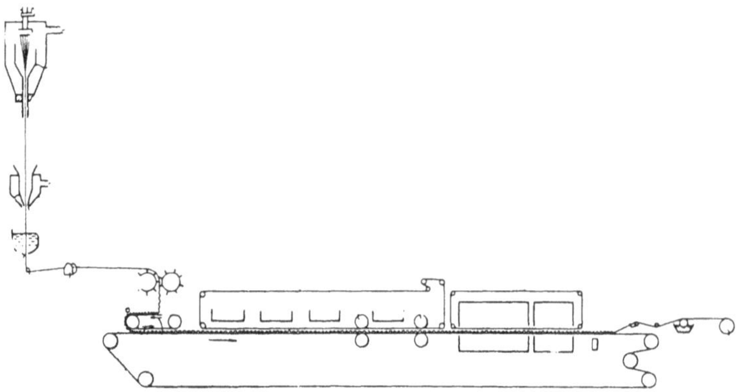

3.26 NP-type spinning apparatus.[17]

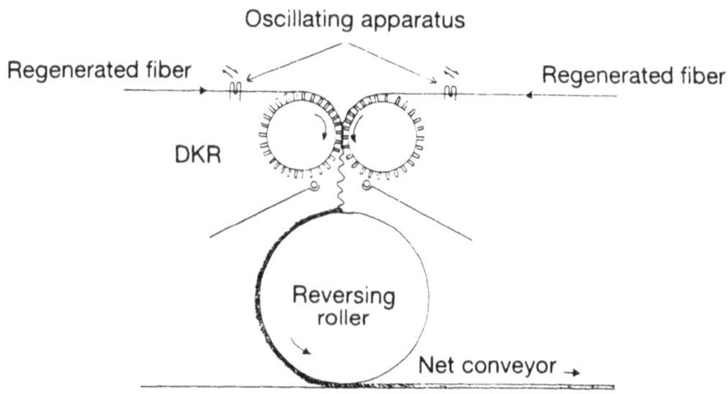

3.27 Apparatus for transferring filaments to the net conveyer.

instrument with an amplitude of several cm just before the DKR, so that the fiber on the roller forms an endless narrow belt with a width of several cm. The fiber belt is transferred from the roller to the net conveyer upside down so that the fiber can be withdrawn from the fiber belt in an orderly fashion. The fiber belt on the main net is treated with sulfuric acid and finishing oil, then dried and finally humidified to a water content of 11%. The fiber belt is covered with a thin net while it runs through these processes to maintain its order.

Figure 3.28 shows the withdrawing apparatus. This apparatus has two functions:

1 To withdraw fiber from the fiber belt and let it run forward.
2 To control the position of the end of the belt where the fiber is drawn out.

Sometimes the fiber is withdrawn from the fiber belt with a short section not undone. This can lead to three effects:

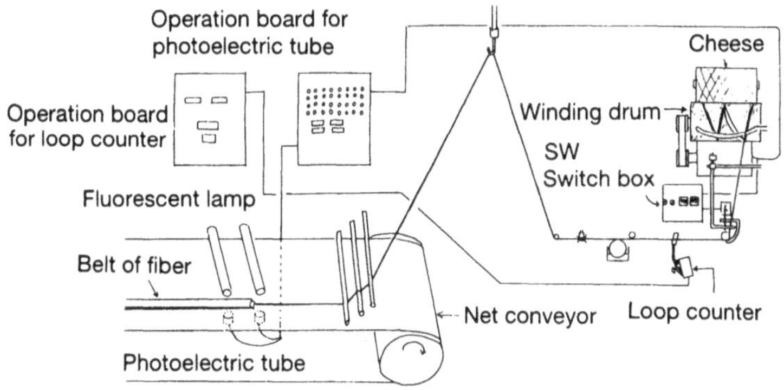

3.28 Apparatus for withdrawing filaments from the net conveyer.

1 Undoing of the belt to a straight fiber.
2 Forming of a knot (or loop).
3 Fiber breakage.

Effects 2 and 3 cause quality problems and reduce throughput. Therefore, preventing these imperfections is an important matter.

The positioning of the end of the belt is controlled as follows. The endpoint of the belt is detected by means of a light sensor system, where a light is set above the main net and a photo cell is set underneath it. If the withdrawal speed is too low, the belt cuts the light beam. The winding speed is raised to a higher level until the photo cell receives the light; once the photo cell receives the light, the winding speed is reduced to a lower level.

There are many reasons for changes in the end point of the belt, such as differences in length and denier of the fiber and difficulty in withdrawing the fiber from the belt.

The merits of employing a net conveyer for the high speed spinning process are:

1 The much lower speed of the net conveyer compared with the spinning speed provides sufficient time for regeneration, drying and humidifying of the fiber.
2 Fiber with high tensile strength and low boil shrinkage can be manufactured since it is treated in the spinning process without tension.
3 No fiber breaks occur on the net. Even if the fiber is broken at winding, an operator can withdraw the fiber from the fiber belt on the net and re-start winding.
4 The investment cost and maintenance charge are reduced because the net speed is low and the number of machine parts is small.
5 It is theoretically possible, using a net conveyer, to realize high-speed

manufacturing at several thousand m/min, if suitable high-speed spinning technology can be developed.

3.3.3.3 Fiber structure and properties of NP fiber

Table 3.4 compares the characteristics of the NP fiber with those of the Hank fiber and the fiber from the continuous Hoffman process. The NP and Hank fibers have high tensile elongation and low boil shrinkage compared with the fiber from the continuous process, because of the treatment under low tension in the regeneration, drying and humidifying processes. The fiber from the continuous process has the smallest number of fluffs per unit fiber length, the second best is the NP fiber. The neps found in the NP fiber stem from loops. From the viewpoint of ease of feeding, fiber properties and fiber quality, the NP fiber is considered to have a good aptitude for the weft of woven fabrics. Recently, the NP fiber has attracted notice because of its high aptitude for the air-jet loom process (Fig. 3.29),[18] introduced mainly for the weaving of regenerated cellulose fiber.

Table 3.4 Fiber properties

Spinning method	Tensile strength (g/d)		Tensile elongation (%)		Shrinkage at the boil (%)	Water absorption (%)	Stiffness (mg/100 filaments)
	Dry	Wet	Dry	Wet			
Hank method	2.0	1.3	14.0	20.0	1.3	86	63
Continuous method	2.7	1.7	10.0	24.0	5.7	84	108
NP method	2.4	1.6	11.0	24.0	3.2	97	16

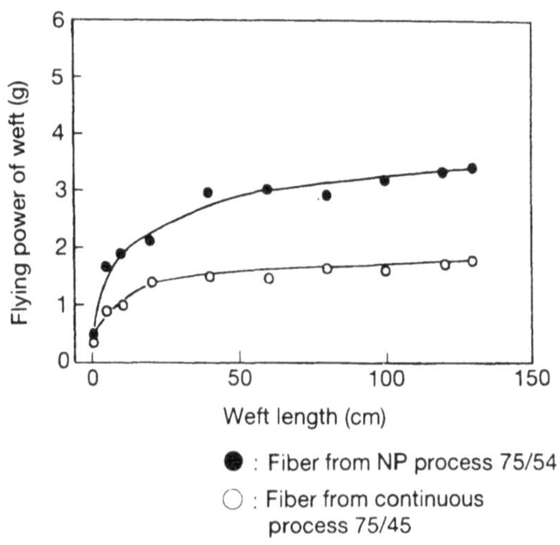

3.29 Dependence of the flying power of weft in an air-jet loom on weft length:[18]

There are drawbacks in the NP fiber such as low abrasion resistance after resin treatment and a tendency to fibrillate. These faults become more apparent as the spinning speed increases, probably due to structural defects in the surface layer as a result of high-speed spinning.

3.3.4 UNP-type manufacturing method

3.3.4.1 High-speed spinning method of the UNP type

The UNP method, an improvement of the NP method, was developed to attain higher productivity and to overcome the drawbacks of the NP method and has the capacity to spin at a speed of 1000 m/min. It employs the same technology as that of the NP method, especially for the processes after spinning. However, in order to attain higher spinning speed and better properties such as abrasion resistance after resin treatment, the following requirements should be fulfilled:

1 Increase of coagulation ability of the spinning funnel.
2 Achievement of high elongation.
3 Optimization of elongation and coagulation.

Requirements 1 and 2 are necessary for high spinning speed and 3 is for improvement of fiber properties.

Figure 3.30 shows the dependence of the abrasion resistance on the

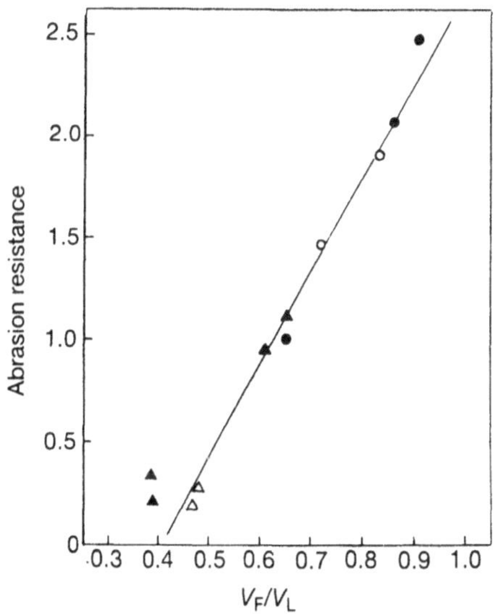

3.30 Dependence of the abrasion resistance of fiber on the ratio of final spinning speed to maximum elongation speed.[19] Abrasion resistance values are relative to values at a final spinning speed of 380 m/min. ● 380m/min, ○ 430 m/min, ▲ 600 m/min, △ 1000 m/min.

ratio of the final spinning speed to the maximum elongation speed due to spinning water, V_F/V_L. Here, the final speed V_F is the speed of the fiber drawn by a DKR and the maximum elongation speed at spinning V_L is the speed of the fiber which is produced only by spinning water free from the force exerted by a DKR, or the speed of the fiber at the point where the elongation by the drawing force of the DKR begins. From Fig. 3.30 it is clear that the abrasion resistance rises as V_F/V_L increases and, in order to improve it, it is necessary to elongate the fiber at an early stage of coagulation.[19]

Figure 3.31 shows that a straight funnel with smaller diameter gives fibers with less fluffs because the higher elongation in the early stage of coagulation improves the balance of elongation to coagulation. However, it has been shown that such a straight funnel causes turbulence of the spinning water and therefore the fiber path was disturbed and even broken. A new type of spinning funnel was introduced in order to solve this problem. This new type of funnel has a tapered end with the smallest diameter at the foot (Fig. 3.32).[20] The aim of this funnel is to prevent breakage of the fiber by turbulent water and at the same time to attain high elongation by increasing the speed of the spinning water.

In addition to this, it was shown that the second funnel, which was important to achieve the necessary coagulation level in the NP method, produced a large resistance against the fiber in the UNP method. Therefore, a second CJ was introduced instead of the second funnel to decrease the resistance from the water and to increase the coagulation

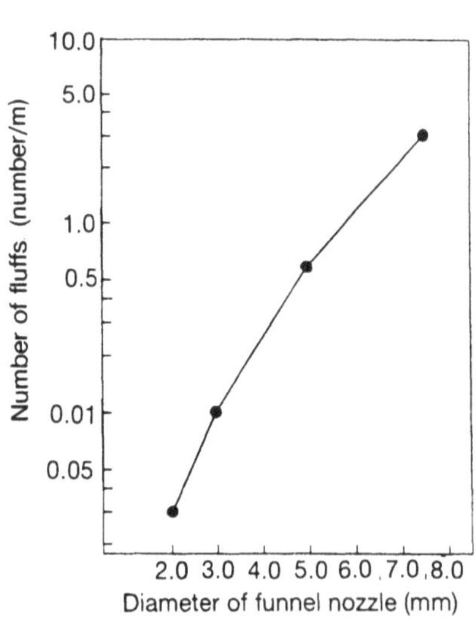

3.31 Relationship between number of neps and the diameter of the spinning-funnel nozzle.[11]

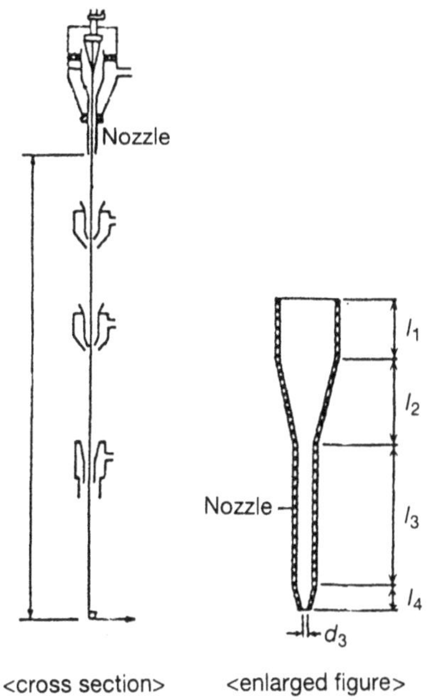

3.32 UNP-type spinning funnel.[20]

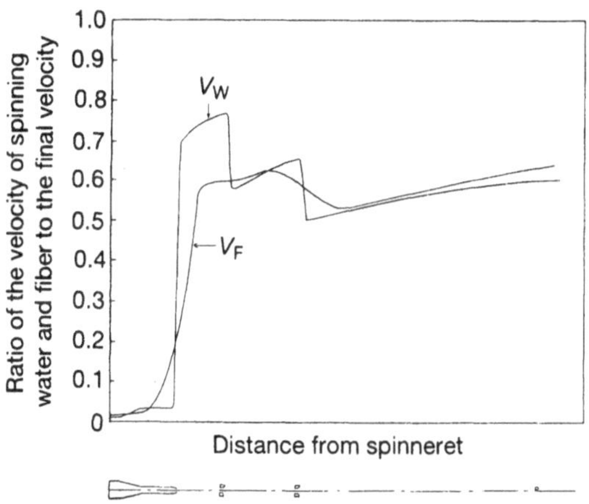

3.33 Change of velocity of spinning water and fiber in UNP-type spinning.[21]

ability. In addition to this, the freefall length was increased. Figure 3.33 shows the change in velocity of the spinning water and the fiber, and the change in degree of coagulation.[21]

As shown in Table 3.5, the employment of this type of spinning

Table 3.5 Fiber properties (high elongation funnel; see Fig 3.32)

		High elongation funnel (UNP)				Conventional (NP)	
Spinning conditions	Spinning speed (m/min)	1000	1100	1200	1500	400	600
	Length of $l_1 + l_2$ (mm)	270	270	270	270	275	190
	Length of l_3 (mm)	240	240	240	240	300	300
	Length of l_4 (mm)	40	40	40	40	–	–
	Diameter of d_3 (mm)	1.4	1.3	1.2	1.0	7.5	7.5
	Length of free fall (mm)	2700	2700	2700	2700	1350	1200
	The 1st coagulant temp. (°C)	50	55	60	65	40	45
	The 1st coagulant vol. (ml/min)	1000	1000	1000	1000	800	1200
	The 2nd coagulant temp. (°C)	80	80	80	80	–	–
	The 2nd coagulant vol. (ml/min)	900	1000	1100	1200	1200	1500
	The 3rd coagulant temp. (°C)	80	80	80	80	–	–
	The 3rd coagulant vol. (ml/min)	1500	1500	1500	1500	–	–
	Tension (g)	10	11	15	20	13	12
Fiber properties	Dry tensile strength (g/d)	2.3	2.3	2.3	2.3	2.3	2.1
	Dry tensile elongation (%)	14.0	13.0	12.2	10.0	13.0	10.3
	Wet tensile strength (g/d)	1.4	1.3	1.3	1.3	1.4	1.2
	Wet tensile elongation (%)	23.0	20.0	18.0	16.0	21.0	17.5
	Fibrillation grade	9.0	8.5	8.0	7.5	7.0	5.0
	Stiffness (mg/100 filaments)	1.2	1.1	1.0	0.9	1.0	0.3
	Fluffs on the surface of the cheese (n)	1.0	1.2	1.5	2.0	2.5	5.0

apparatus has made it possible to manufacture fibers with low fluff level, good abrasion resistance and low fibrillation tendency.

3.3.4.2 Other characteristics of the UNP process

The frequency of loops in the UNP process is one tenth of the NP process. This is considered to be because entanglements in the fiber belts occur less often using DKR drawing and it becomes easier to withdraw the fiber from the fiber belts at high speeds. The effect is considered to stem from the fact that fiber drawn by the DKR reaches the net conveyer before the crimp produced by the blades of the DKR disappears.

3.3.5 Summary

High-speed spinning of Bemberg fiber at 1000 m/min has been developed by improvement of the stretch-spinning method, first developed by Thiele in 1901, and development of the net conveyer method. This process also improves fiber properties such as tensile strength, abrasion resistance and fibrillation tendency. As the net process does not restrict spinning speed, further increases in spinning speed will be achieved by improving the coagulation and elongation technology in the funnel.

3.4 Spinning technology of spandex fibers

3.4.1 Manufacturing method of spandex fiber

Research into spandex polyurethane fiber started in the 1930s in Germany. In the 1940s, the IG company developed Perlon U, which however was a non-elastic polyurethane fiber rather like nylon. The elastic polyurethane fiber 'Lycra' was developed in 1959 by Du Pont, and thereafter many companies started developing elastic polyurethane fibers.

The segmented polyurethane used for elastic polyurethane fibers is a copolymer of a soft segment and a hard segment. Rubber-like elasticity is given by combining a soft segment which gives polymer elongation and a hard segment which gives polymer strength.

In the polymerization process, a high molecular-weight diol is converted to a prepolymer which has an isocyanate group at each end by combining it with two molar equivalents of diisocyanate. Polytetramethyleneglycol (PTMG), a polyadipate, or polycaprolactone are used as the high molecular weight diol and diphenylmethane 4,4′-diisocyanate (MDI) or toluene-2,4-diisocyanate (TDI) is used as the capping agent.

This prepolymer is then converted to high molecular weight polyurethane by combining it with a chain propagation agent such as a diamine or some other bifunctional active hydrogen compound such as a diol. The urea bonds (in the case of diamine) or urethane bonds (in the case of diol), which form in this reaction, produce hard segments. Hydrazine and ethylene diamine are the most commonly used diamines.[21]

As there is a large amount of free diisocyanate in the pre-polymer, the polymer produced has the following structure.

—IAIAIAI—I-I-I—IAIAIAIAI—I-I-I—IAIAIAI—

Here I is diisocyanate, A is diamine and — is high molecular weight diol. The sequence IAIAIAIAI corresponds to a hard segment and —I—I—I— corresponds to a soft segment.

Generally, a dry- or wet-spinning method is used for manufacturing spandex fiber but the melt-spinning method is also applied. There are two methods of wet spinning, one using a solution of the final polymer and the other a prepolymer. In the former method, the polymer solution is extruded into a coagulation bath through a spinneret and after coagulation, combination and fusion between filaments take place, the fiber is wound up. When using prepolymer, the solution of prepolymer is extruded into diamine solution where chain propagation takes place. Therefore this spinning method is often called reaction spinning (Fig. 3.34).[22]

In the case of dry spinning the heated polymer solution is extruded from a spinneret into a hot gaseous atmosphere in a spinning tube, where the solvent is removed. The filaments are combined to form a yarn, fused

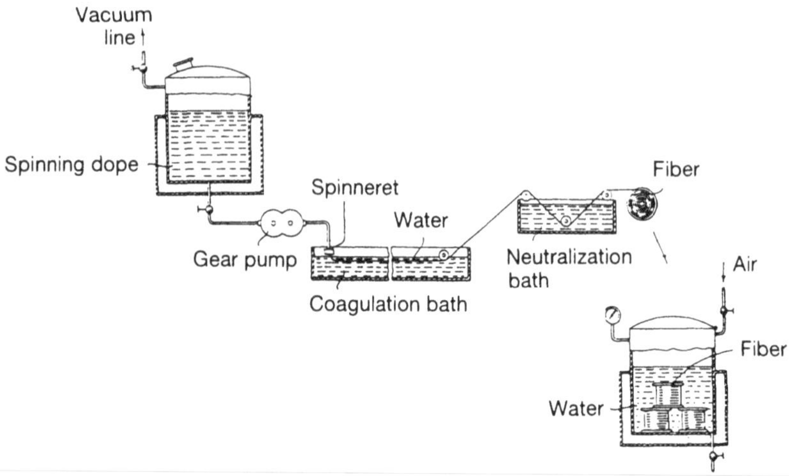

3.34 Wet-spinning method for spandex.[22]

with each other and wound up. This is now the most usual spinning method employed for spandex manufacture.

3.4.2 Dry spinning of spandex fibers

3.4.2.1 Analysis of the spinning process

The process of structural formation of the fiber in dry spinning is more complex than in melt spinning, since the former is a two-component system, while the latter is a one-component system. However, assuming that there is no change of phase due to vaporization of solvent and that the fiber can be treated as a continuous body, the spinning process can be analyzed theoretically like the melt-spinning process.[23]

First, cylindrical co-ordinates $(r.z)$ are set up as denoted in Fig. 3.35. The region surrounded by two horizontal planes which are separated by a vertical distance Δz and a cylindrical surface of the fiber is considered. For this region, four equations define the balance of materials (polymer and solvent), momentum and energy. By transforming these equations, four independent equations can be introduced that define the weight ratio of solvent, sectional area of fiber, spinning tension and fiber temperature, and express the spinning conditions. Figure 3.36 shows the results of calculations using these equations. In these figures the dotted lines denote the case (a) that solvent vaporization is restricted by a boundary layer and the continuous lines denote the case (b) that it is restricted by diffusion (cf. Fig. 3.37).

Figure 3.36(a) shows that the solvent concentration in the region near the spinneret is higher for case (a) than for case (b) but the concentration decreases rapidly in the next region, the concentration for case (a) becomes lower than that of case (b) and the time until the solvent is

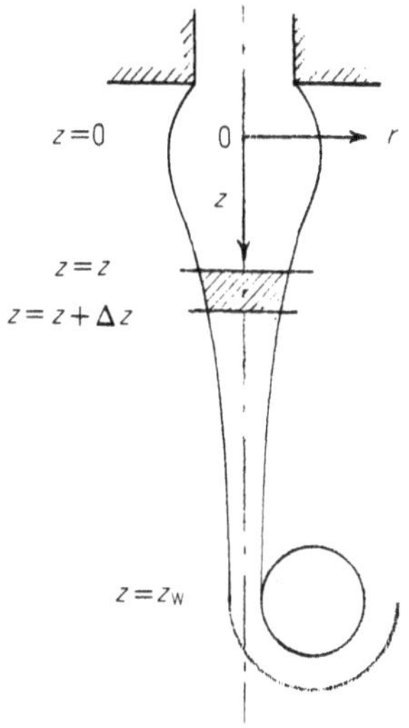

3.35 Schematic diagram of spandex extrusion process.

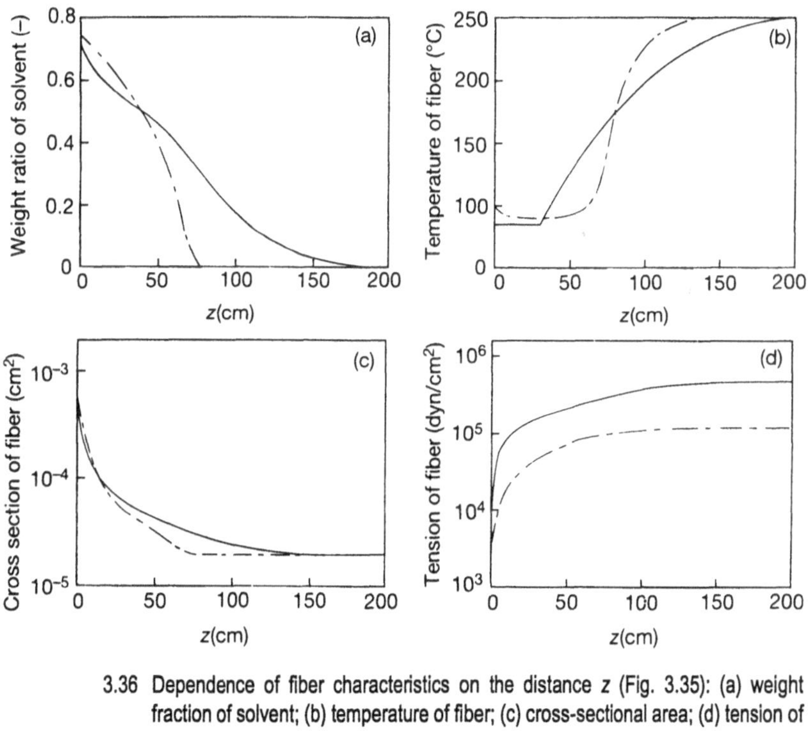

3.36 Dependence of fiber characteristics on the distance z (Fig. 3.35): (a) weight fraction of solvent; (b) temperature of fiber; (c) cross-sectional area; (d) tension of fiber.

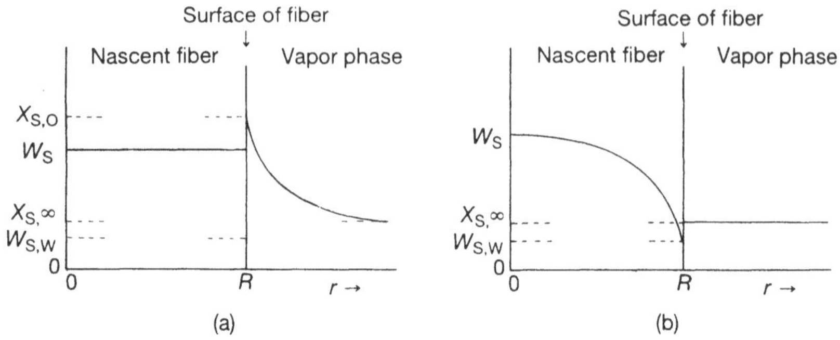

3.37 Radial distribution of the concentration of solvent in the fiber. (a) Solvent vaporization restricted by a boundary layer. (b) Solvent vaporization restricted by diffusion.

completely removed is longer for case (a). In this figure the concentration dependence of the diffusion coefficient is not taken into consideration. In fact, the diffusion coefficient is considered to decrease as the solvent concentration decreases, so it takes much longer to remove the solvent completely.

Figure 3.36(b) shows that the temperature of the fiber drops at first, then remains constant for a while and gradually rises to the temperature of the circumferential gas. The first drop is the result of the vaporization of the solvent which takes latent heat from the fiber. Then the temperature reaches wet-bulb temperature and remains there because the latent heat equals the heat transferred from the circumferential gas to the fiber. Thereafter, the temperature starts to rise as the amount being vaporized decreases.

The cross sectional area (Fig. 3.36(c)) drops rapidly at first because of elongation and then decreases more gradually because of the vaporization of the solvent. Thereafter, as the amount of solvent vaporization decreases and the fiber becomes solid, the change of cross-sectional area diminishes.

The tension of the fiber (Fig. 3.36(d)), which is very small at first, rapidly increases as the solvent vaporizes, but soon reaches a plateau.

These four values shift by a factor of two along the horizontal axis if the amount of spinning dope increases two-fold. This means that if the amount of spinning dope is increased in order to increase the spinning speed, the length of the spinning tube should be extended proportionately to the increase in the amount of dope.

3.4.2.2 Practical dry spinning of spandex

Figure 3.38 shows the conceptual diagram of a spinning apparatus for dry spinning of spandex. The polymer solution, which is already

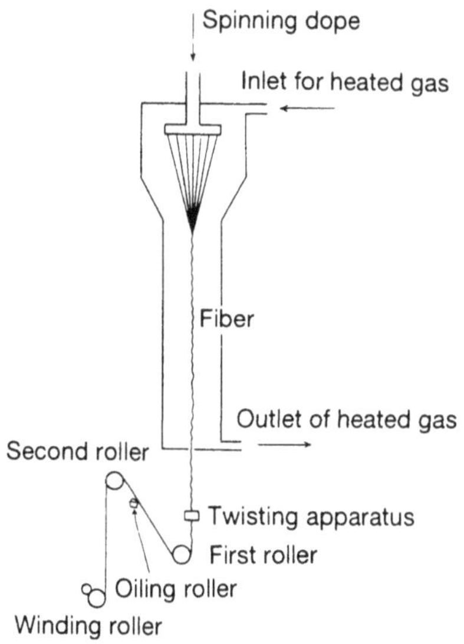

3.38 Schematic diagram of dry-spinning apparatus.

de-aerated, filtered and heated up to the required temperature, is extruded from a spinneret into a spinning tube. From the top of the tube, inert gas such as Kempf gas is introduced into the tube. The solvent is removed from the fiber, which becomes thinner as it passes down the tube. After emerging from the tube, the fiber is twisted. The twist travels upstream along the fiber to a point in the upper part of the tube where the filaments comprising the fiber fuse together and form an aggregate of filaments. This fusion is indispensible for subsequent processing of the fiber if it is not to be damaged. Various twisting devices such as air jet, liquid flow or roller can be used. The twist imparted is removed by the time the fiber reaches the first roller by means of the tension of the fiber itself. The fiber then receives finishing oil by contact with the finishing roller and is wound up by way of the second roller.

Various technological improvements have been made to obtain stable spinning and diminish denier fluctuation. Figure 3.39 shows a technological improvement concerning the flow control of the heated gas.[24] The heated gas flows down through a ring-shaped passage formed between the spinning tube and a cylindrical body set under the spinning nozzle and is removed from the tube at a point sufficiently far from the nozzle. This flow control prevents the oscillation of filaments and minimizes denier fluctuation.

Generally, in dry spinning the heated gas is supplied in parallel with the direction of the fiber either co-currently or counter-currently, but Fig. 3.40 shows a method where the heated gas is supplied perpendicularly to

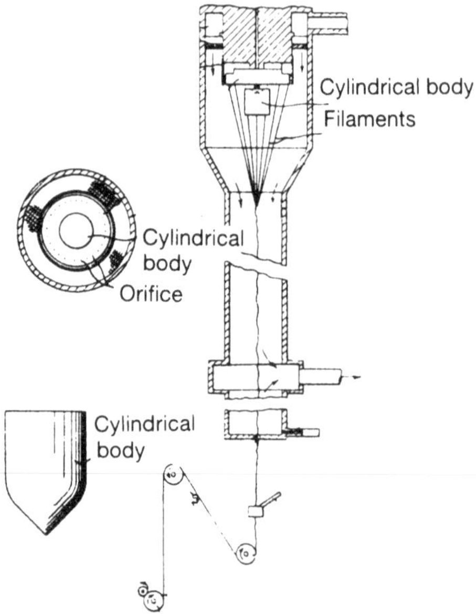

3.39 Flow control method for heated gas.[24]

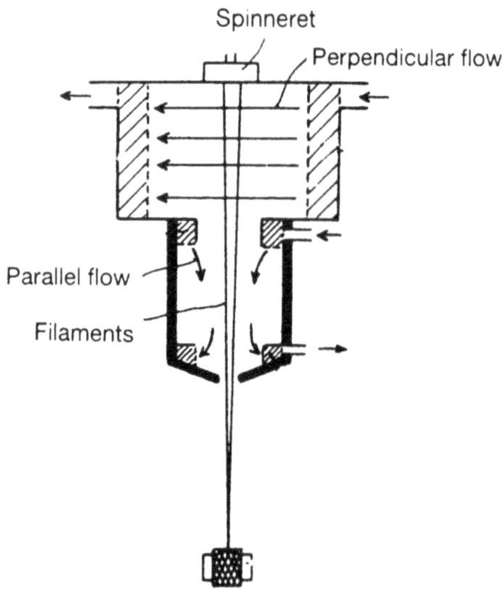

3.40 Flow of heated gas: cross-flow system.[25]

the fiber as in melt spinning.[25] The velocity of the gas is 1–30 cm/sec. This method prevents gas turbulence and heat transfer fluctuations and therefore decreases the entanglement of fibers and denier fluctuations. Moreover, this method accelerates the vaporization of solvent and facilitates higher spinning speed.

3.4.3 High-speed spinning of spandex

The spinning speed of spandex is typically about 500 m/min; much slower than that of synthetic fibers such as PET made by a melt spinning method. In the spinning of spandex, an increase of the fiber tension causes undesirable changes such as an increase of modulus and a decrease of tensile elongation. Therefore, lowering the fiber tension is the main theme of the technological development in order to attain high-speed spinning. It is also necessary to increase the rate of solvent vaporization.

The developments required are:

1. A method to supply the heat energy needed to evaporate the solvent. The amount of solvent to be evaporated increases proportionally to the increase of spinning speed, yet it is not desirable to lengthen the spinning tube proportionately from the viewpoint of economy and fiber tension. Therefore, the residence time of the fiber in the tube must become shorter. An increase of the gas supply rate causes turbulent flow. An increase in temperature of the gas is limited by the melting point of the fiber.
2. Technology for diminishing fiber tension. As mentioned before, an increase in the fiber tension causes a rise of modulus and a decrese of tensile elongation. The velocity of the gas becomes lower than that of the fiber. The resistance of the gas to the fiber increases as the fiber velocity increases.
3. High-speed twisting technology. As mentioned before, the fiber is twisted after it comes out of the tube by means of an air jet, liquid flow or rotating roller. The rate of revolution of the twister must be increased to attain higher spinning speed.
4. High-speed winding technology. It is difficult to wind up spandex yarn as it has low modulus and high tensile elongation. The technological difficulties increase proportionately to the increase of spinning speed.

Research and development is advancing on each of these themes in order to improve fiber properties. However, it is not yet possible to attain high spinning speed because of the restriction that spandex must be spun under low tension to guarantee the low modulus of the final product.

References

1 Kamide K, *Thermodynamics of Polymer Solutions*, Elsevier, Amsterdam, p. 455, 1990.
2 Kamide K and Manabe S, *ACS Symposium Series*, **269**, 197, 1985.
3 Kamide K, *Thermodynamics of Polymer Solutions*, Elsevier, Amsterdam, p. 582, 1990.
4 Ashai Chemical Ind., Jap. Pat. (Examined) 1967-4293; US Pat. 3, 384,694.
5 Jap. Pat. (Examined) 1965-26212.
6 Jap. Pat. (Examined) 1967-2014.
7 Mitsubishi Rayon, Jap. Pat. (Laid open) 1978-81723.
8 Ashai Chemical Ind., Jap. Pat. (Laid open) 1976-35716.
9 Egami O, private communication.
10 Meyer K H, *Hochpolymere Chemie*, **bd II**, Die Hochpolymere Verbindungen, Academisches Verlagsges, Leipzig, p. 274, 1940.
11 DRP 154507, 157157, 178889, 199772.
12 Oka E, Munekata E and Watano M, *Chemical Fibres*, Maruzen, 1956.
13 Bozza G, 'Il process di filatura del cuprammoniacale'.
14 Elsasser V, *Koll.Z.*, **111**, 174, 1948; **112**, 120, 1949; **113**, 37, 1949.
15 Jap. Pat. (Examined) 1969-22204.
16 Ogawa, T, private communication.
17 Ashai Chemical Ind., Jap. Pat. (Examined) 1972-29926; US Pat. 3,689,620.
18 Iwase, K, Lecture to symposium of Japan Fibre Society, 1985.
19 Iwase, K and Moroe I, private communication.
20 Ashai Chemical Ind., Jap. Pat. (Examined) 1991-72721.
21 *Manufacture, Structure and Properties of Fibers*, Japan Fibre Engineering Society, p. 58, 1983.
22 Jap.Pat. (Examined) 1961-23808.
23 The Society of Fiber Science and Technology, Japan (Ed), Formation of Fibers and Development of their Structure – III, Kagaku-dojin, Kyoto, 1971.
24 Jap. Pat. (Examined) 1964-16667.
25 Toyobo, Jap. Pat. (Examined) 1970-30045.

4

Spinning for nonwovens

Ei-ichi Kubo
Unitika Ltd, Uji, Japan
Masaharu Watanabe
Unitika Ltd, Okazaki, Japan

4.1 Introduction

The production of nonwovens has increased remarkably in recent years. The annual worldwide production is now over 1.2 million tons and the annual Japanese nonwovens production[1] reached about 0.16 million tons in 1989. There is a wide range of uses of nonwovens, ranging from agriculture, civil engineering, construction, automotive and industrial materials to medical/surgical, sanitary and household products. There are also many processes for making them. These include dry-laid, wet-laid, spunbonded, meltblown, flashspun and spunlaced nonwoven processes.

Each manufacturing process causes the nonwovens to have certain specific characteristics, and each product is used in many applications, particularly suited to these characteristics. The spunbonded process in the narrow meaning of the term (which is the meaning used here) is a manufacturing method comprising three stages in an integrated operation: melt-spinning thermoplastic high polymer, collecting continuous filaments on a moving collection belt to form webs, and binding the filaments in the webs to produce nonwovens. As this process has a high productivity and provides the nonwovens with excellent mechanical properties, plans for expansion of production facilities for this process are announced every year in the nonwovens industry. Meltblown and flashspun processes could also be called spunbonded processes in the wider meaning of that term because these processes consist of a one-step process from spinning to consolidation as in the spunbonded process described above. However, meltblown and flashspun processes differ from the spunbonded process in the nature of the spinning stage in which the polymers are attenuated to form filaments. This means that the properties and form of the filaments made by each process differ and consequently that the nonwoven fabrics from each process differ greatly from each other in their properties.

The flashspun process is characteristically different in the spinning stage from the other two processes. In spunbonded and meltblown processes, molten polymer from an extruder is spun into continuous filaments or fibers, and so the spinning stage is a melt-spinning. The spinning stage of the flashspun process, on the other hand, is not melt-spinning. In the flashspun process, the filaments are obtained from a polymer solution of an organic solvent which has a low boiling point, is inactive to the polymer, and does not dissolve the polymer at room temperature but does dissolve it at a high temperature and pressure. A mixture of polymer and solvent is brought up to a high temperature and pressure by heating to obtain a solution. The solution obtained is flashed through a small orifice into the atmosphere at room temperature to form filaments, and is therefore a type of dry-spinning. In the spunbonded and meltblown processes and in normal dry-spinning, only one filament issues from each orifice. However, flash-spinning is distinctive. It enables the polymer solution to form many filaments from one orifice. These filaments are termed plexifilaments and form a three-dimensional structure. This chapter concentrates on the melt-spinning methods, particularly on the spunbonded and meltblown processes.

4.2 Spunbonded process

The spunbonded process was first commercialized by Du Pont in 1959 when the company started selling PET spunbonded nonwovens under the trade name 'Reemay'. Figure 4.1, a schematic representation of a spunbonded process commonly used by many manufacturers, shows the filaments being withdrawn by air-jets, but the filaments can alternatively be withdrawn by rolls instead of air-jets. There are two types of spinneret in commercial use; one is circular and the other rectangular. There are also two types of air-jet, circular and rectangular. Figure 4.2 presents some examples of spinning systems disclosed in the patents. The most important point in producing nonwovens is opening and arranging the filaments into uniform webs. There is no definite combination of the spinneret and the withdrawal device to achieve a uniform web. The producers have developed many technologies to improve filament opening suitable for making webs with good uniformity of appearance: air diffusers fitted at the outlets of air-jets, static electricity generated by rubbing the filaments on metal plates, corona charging by passing the filaments through a corona discharge device, and traverse devices, for example. Any filament-forming polymer can be used in the spunbonded process and for making spunbonded fabrics. Among these polymers are polyethylene terephthalate (PET), polyolefines, polyphenylene sulphide (PPS), polybutylene terephthalate (PBT) and polyamides. The spinning speeds are generally between 3000 and 6000 m/min. The selection of polymers depends on the end-uses. Combinations of polymers are also adopted to meet the needs of varied applications and nonwovens with

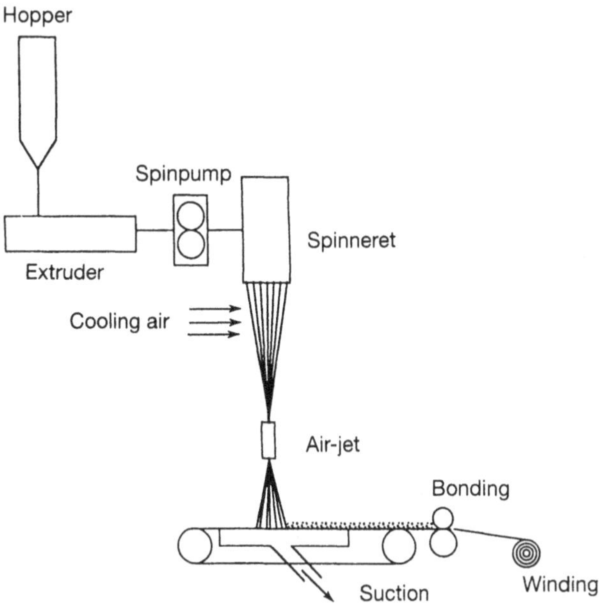

4.1 Spunbonded process.

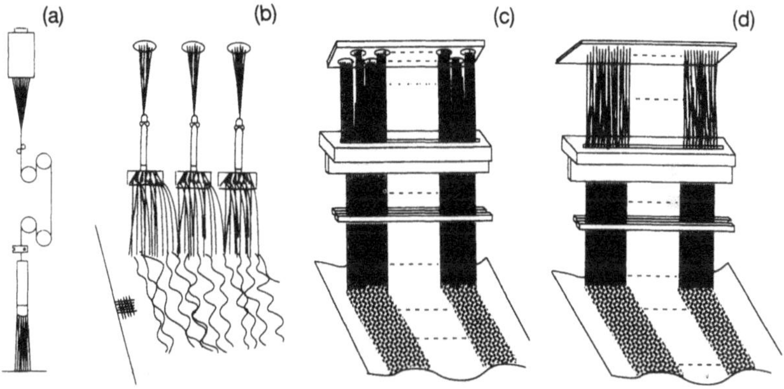

4.2 Spinning systems for spunbonded nonwovens. (a) Reference 2, (b) Reference 3, (c) Reference 4, (d) Reference 5.

high performance. For example, there is Unitika's spunbonded fabric under the trade name 'ELEVES®' which has excellent strength and softness. This spunbonded fabric is made of PET and polyethylene (PE) and makes use of the characteristics of the two components: the high strength of PET and the softness of PE. 'ELEVES®' consists of continuous bicomponent filaments having a core component made of PET and a sheath component made of PE. Some producers are making new laminates of spunbonded fabrics consisting of different polymers. One of these products is Unitika's 'ELFIT', which was first publicised in April 1991, which is a laminate of two kinds of spunbonded fabric. One side of the laminate is PET spunbonded fabric and the other side is 'ELEVES®' consisting of bicomponent filaments with PET core and PE sheath. The polymers on each side have different melting temperatures. Thus 'ELFIT' improves the processability in combination with other materials, making it easier to produce composites and to meet the versatile needs of nonwovens.

There are nonwovens made of ultra-fine filaments with a fineness of 0.5 denier or less. These nonwovens are produced from webs consisting of multicomponent filaments such as the sea–island type or the dividable type which are composed of two kinds of polymers immiscible with each other. Nonwovens made of ultra-fine filaments are created either by dividing multicomponent filaments in raw nonwoven form after consolidating the webs or by dividing the filaments and consolidating the webs at the same time. The division technique used with the sea–island type is the removal of the sea section polymer in a solvent. After this dissolution, the island portion is left behind to create ultra-fine filaments. This method is a complicated process because of the necessity to remove the sea portion, but this technique now provides the finest filaments. The technique was first developed by Toray. Researchers at the Okamoto laboratory have already developed the finest PET fiber with a fineness of 0.00009 denier and a diameter of 0.1 μm. They claim to produce the finest in the world. Figure 4.3[6] shows multicomponent filaments of the sea–island type. The division technique for the dividable

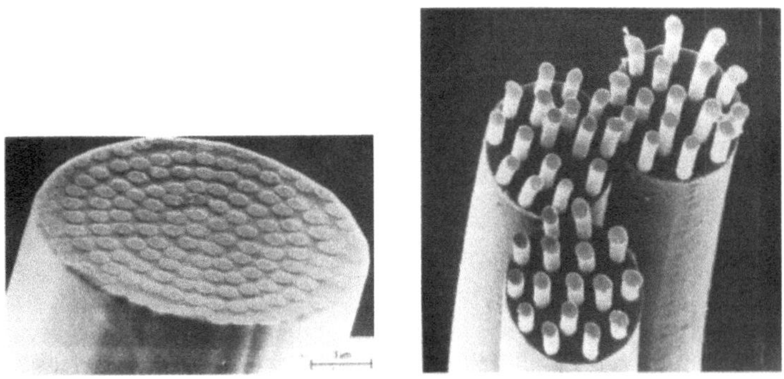

(a) (b)

4.3 Multi-component filaments of islands-in-a-sea-type. (a) Filament section. (b) The island portions are left to create the ultra-fine filaments after the dissolution of the sea polymer.

type is separation of the two polymers forming the filament by use of mechanical force, thermal action or impact generated by highly pressurized water. The division method with highly pressurised water[7] causes separation and entanglement of the filaments at the same time to consolidate the web into an end-product consisting of ultra-fine filaments; this method is well worth noting. Figure 4.4[8] shows multicomponent filaments of the dividable type.

Among polyolefine spunbonded fabrics, polypropylene (PP) is the most widely used. PE spunbonded fabric is also produced but the output is now very small. Among the PE resins, linear low density polyethylene (LLDPE) has good spinnability. LLDPE can be easily melt-spun into filaments at a high speed, over 3500 m/min, in conventional equipment by conventional techniques. It provides a spunbonded fabric consisting of filaments having a fineness of 2–3 denier. The tensile strength of LLDPE spunbonded fabric is inferior to that of PP spunbonded fabric but it exhibits an exceptional softness unattainable by PP. A higher spinning speed can be obtained by using a blend of low density polyethylene (LDPE) and PP.[9] In the patent examples, a series of blends were formed from LDPE having a density of 0.9096 g/cm^3, a melting temperature of 103.4°C, and a melt index of 70 and PP having a density of 0.9022 g/cm^3 a melting temperature of 164.0°C, a melt flow rate of 8.7 and a Q value of 8.8 (ratio of weight-to-number average molecular weight). Figure 4.5, which presents results obtained using conventional spinning equipment, depicts the maximum melt-spinning speed at 260°C of various blends of LDPE and PP. The filaments were melt-spun using a spinneret plate having 70 holes of circular cross-section at a polymer flow rate per hole of 0.90 g/min, and were withdrawn by a roll. LDPE blended with a small

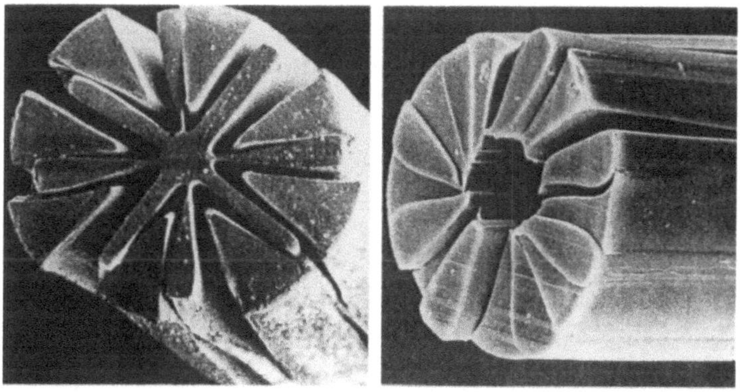

(a) (b)

4.4 Multi-component filaments of dividable type. (a) The filament with triangular section is PET and the other filament is polyamide. (b) The filament with a rectangular section is PET and the other is polyamide.

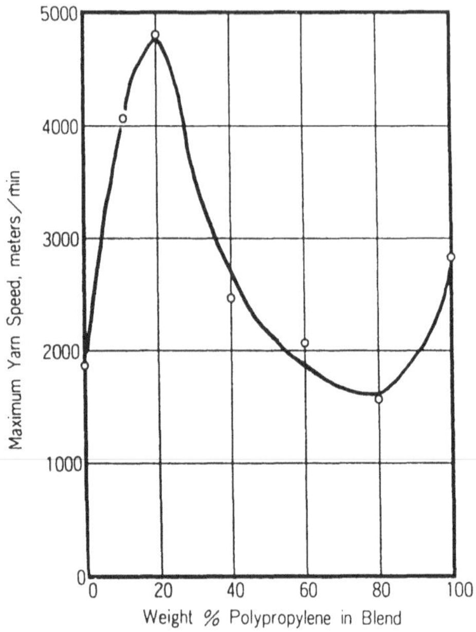

4.5 Speed of filament breakage when spinning LDPE blended with PP.

amount of PP increases the spinning speed. Analysis of the filaments melt-spun from blends at higher speeds showed that PP dispersed in LDPE takes the form of droplets in the filaments, mostly about 5μm in diameter by about 7μm in length. It is believed that PP droplets in LDPE enhance the crystallization rate of LDPE and consequently the blend polymers can be melt-spun at much higher speeds than unblended LDPE. On the other hand, blending PP with a small amount of LDPE reduces the spinning speed relative to unblended PP. It is also believed that LDPE in PP hinders the crystallization of PP and so the blend detrimentally affects the spinning speed. This patent is an interesting example of melt-spinning of polymer blends.

4.3 Meltblown process

The meltblown process was initiated in 1951 by the US Naval Research Laboratories in an effort to develop organic microfibers of less than 1μm in diameter. The researchers aimed to produce fibers which were to be used for microdenier filters to collect radioactive particles in the upper atmosphere. This initial research and development work on meltblown processes was continued by an Exxon affiliate in the mid 1960s, and Exxon successfully demonstrated the first meltblowing unit for producing microdenier webs in the early 1970s. Figure 4.6 is a schematic

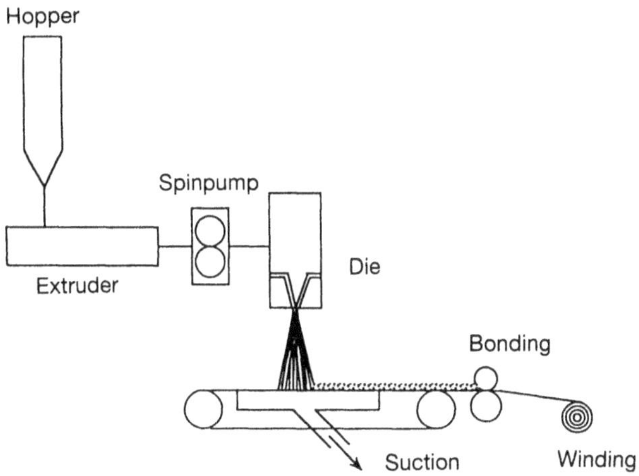

4.6 Meltblown process.

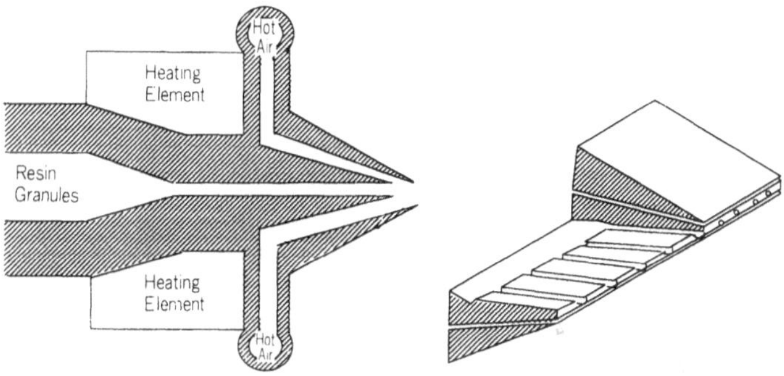

4.7 Meltblowing die (Exxon type).

representation of a meltblown process. The structure of the die used in meltblown processes differs greatly from that commonly used in spunbonded processes. In meltblown processes the distance from die to conveyor is very short, i.e. 20–50 cm, compared to 100–200 cm in spunbonded processes. This is a process in which high velocity gas blows molten polymer to form fine fibers. The die has a row of small holes of a diameter of less than 1 mm, which are 0.5 mm apart, center to center. Along the row of small holes, the die also has slits through which gas, heated to a temperature above 300°C, blows on to molten thermoplastic polymer coming out of each hole. Generally the heated gas is air. Figure 4.7 shows the meltblowing die designed by Exxon. Many different polymers have been used in the meltblown process, most commonly PP, but also LDPE, high density polyethylene (HDPE), PET, polyamides, polystyrene, polyurethanes, PBT and PPS.

The meltblown process consumes a large amount of heated gas to obtain microdenier fibers. The consumption of gas depends on the polymer, the molecular weight of the polymer and the polymer flow rate per hole. This process normally consumes forty to fifty times as much air by weight as the polymer flow rate and sometimes consumes a hundred times. Meltblown processes therefore require a large amount of energy. The force of gas attenuates the fiber at a very high speed. For example, when fine fiber with a diameter of 3μm is being produced at a polymer flow rate of 0.2 g/min per hole by using PP with a density of 0.900 g/cm^3, the spinning speed is calculated to reach around 31 000 m/min, about Mach 1.5. It is necessary to decrease the temperature and consumption of gas to produce fine fibers at a lower energy. This means that it is advantageous in terms of energy consumption to use a polymer with a lower melting temperature, or with a higher flow index, or with both a lower melting temperature and a higher flow index. However, the fiber must also have the properties required for its practical use and this sets a limit to the choice. Shambaugh[10] has reported a study of energy consumption in PP meltblowing, which can be summarized as follows:

1 The attenuation of fiber is mainly dependent on the gas kinetic energy.
2 Most of the gas kinetic energy is wasted without being effectively utilized when coarse denier fibers are obtained even though a sufficient amount of gas is supplied.
3 The actual fibers have a wide diameter distribution, but the diameter distribution should ideally be controlled to be monodisperse.

When a polymer is being attenuated into a fiber, the fiber kinetic energy E (erg) can be expressed as:

$$E = 1/2MV^2 = 8M^3/(\rho^2 \pi^2 d^4) \qquad [4.1]$$

where M is polymer flow rate (g/sec), V is fiber velocity (cm/sec), ρ is polymer density (g/cm^3) and d is fiber diameter (cm). Shambaugh calculated the fiber kinetic energy for different fiber distributions with the same average fiber diameter (10μm) using the above equation (4.1).

The comparisons are as follows:

$$E_1 : E_2 : E_3 = 1 : 3.8 : 27.7 \qquad [4.2]$$

where E_1 is the energy for a monodisperse distribution with a fiber diameter of 10μm, E_2 is the energy for a bimodal distribution of a 50:50 mixture (by number of fibers) of 3μm and 17μm and E_3 is the energy for a bimodal distribution of a 50:50 mixture (by number of fibers) of 1μm and 19μm. As the actual fiber distributions are not so simple as these examples, the above comparison data are only approximate. However, the data indicate that a monodisperse fiber distribution requires much less energy than a polydisperse distribution with the same average fiber diameter. It is a great disadvantage to energy conservation to produce finer fibers than are necessary for the practical use. It can easily be understood without calculating a more complicated equation, that from the equation

$$E = 8M^3/(\rho^2 \pi^2 d^4) \quad [4.3]$$

the energy E increases when d decreases at a constant value of M. It is necessary to equalize the polymer and the gas flow rates at each hole to get a monodisperse distribution. A re-design of the meltblowing die[11] has been carried out by Nippon Kodoshi to improve the uniformity in the polymer and the gas delivery. This die is designed for the heated gas to blow through a labyrinth. Figure 4.8 is the revised meltblowing die and the numeral 26 in the illustration is the labyrinth.

As in the case described in the section on spunbonded processes, the lamination of meltblown fabrics with other materials is employed to improve the nonwoven properties. An example is the lamination of meltblown fabrics with spunbonded fabrics. Figure 4.9 shows the S–M–S process[12] producing a three-ply system from a meltblown fabric sandwiched between two spunbonded fabrics. Other meltblown dies disclosed in the patents include a die capable of spunbonded spinning and meltblown spinning at the same time[13] and a die capable of spinning sheath–core or side-by-side bicomponent fiber.[14]

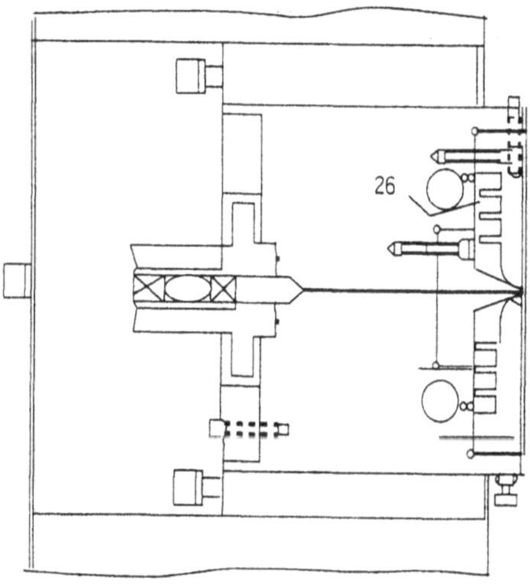

4.8 Meltblowing die for improved uniformity.

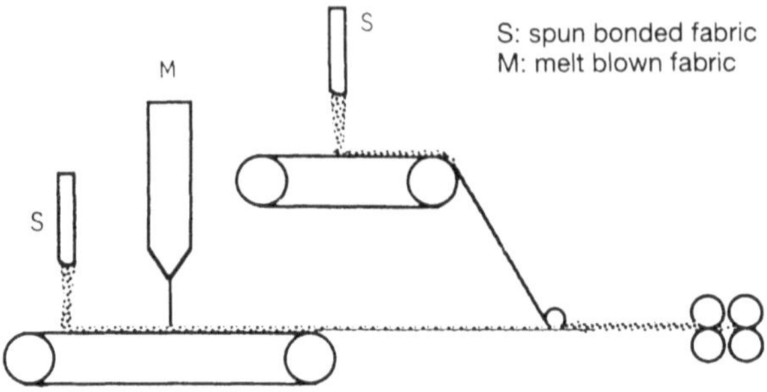

4.9 S–M–S process.

4.4 Conclusions

The main melt-spinning methods for nonwovens have been described. It can be said that Japan leads the world in the melt-spinning technology. The researchers and the producers will continue to create products with higher performance by developing new technologies in combination with polymers and processes.

References

1 Japan Nonwoven Report, Jan 1991.
2 Du Pont, US Pat. 3,338,992.
3 Asahi Chemical Ind., Jap. Pat. (Examined), 1989-15,615.
4 Asahi Chemical Ind., Jap. Pat. (Laid open), 1988-282,350.
5 Asahi Chemical Ind., Jap. Pat. (Laid open), 1988-282,350.
6 Nikkei New Materials, No. 46, No. 50, 1988.
7 Toray, Jap. Pat. (Examined), 1989-47,585.
8 Nikkei New Materials, No. 52, 1988.
9 Du Pont, US Pat. 4,632,861.
10 Shambaugh RL, *Ind. Eng. Chem. Res.*, **27**, 2363, 1988.
11 Nonwovens World, **4** (5), 30, 1989.
12 Kimberley-Clark Corp., US Pat. 4,810,571.
13 Jap. Pat. (Laid open), 1973-99,411.
14 Minnesota Mining and Mfg Co., Jap. Pat. (Laid open), 1985-99,057; US Pat. 4,729,371.

5

The spinning of highly aesthetic fibers

Masao Matsui
Kanebo Ltd, Osaka, Japan

5.1 The evolution of aesthetic fibers

The aesthetic fibers discussed in this chapter are highly fashionable materials which have superior aesthetic and sensual factors such as appearance, colour, handle and touch, softness, bulkiness, and special texture. Highly aesthetic thin fabrics are used for women's dresses and blouses. The medium and thick fabrics are used for suits, skirts, slacks, formal wear, coats and sports/casual wear.

Silk fabrics have been used for women's dresses and blouses for hundreds of years. Recently, much refined silk-like synthetic fibers and fabrics produced by highly advanced technologies have been widely used. The evolution of silk-like fibers and fabrics is shown in Fig. 5.1.

The first generation silk-like fibers, which appeared in the 1960s, have a triangular cross-section and a luster similar to silk. The non-circular cross-sections give the fiber not only a different luster, but also a remarkable change of bending stiffness (bending moment), coefficient of friction, softness and handle. A regular triangular cross-section gives, for example, a fiber with 1.2 times larger bending stiffness compared with that of a round section. The fibers with flat cross-section, on the other hand, have much lower bending stiffness and softer handle.

The handle (hand values) can be measured objectively by Kawabata's Evaluation System (KES), developed recently.[1] The hand values are expressed by various factors such as:

'Koshi' (stiffness).
'Hari' (spread and anti-drape).
'Fukurami' (fullness and softness).
'Shari' (crispness).
'Kishimi' (scroopy feeling).
'Numeri' (smoothness).
'Shinayakasa' (flexibility with soft feeling).

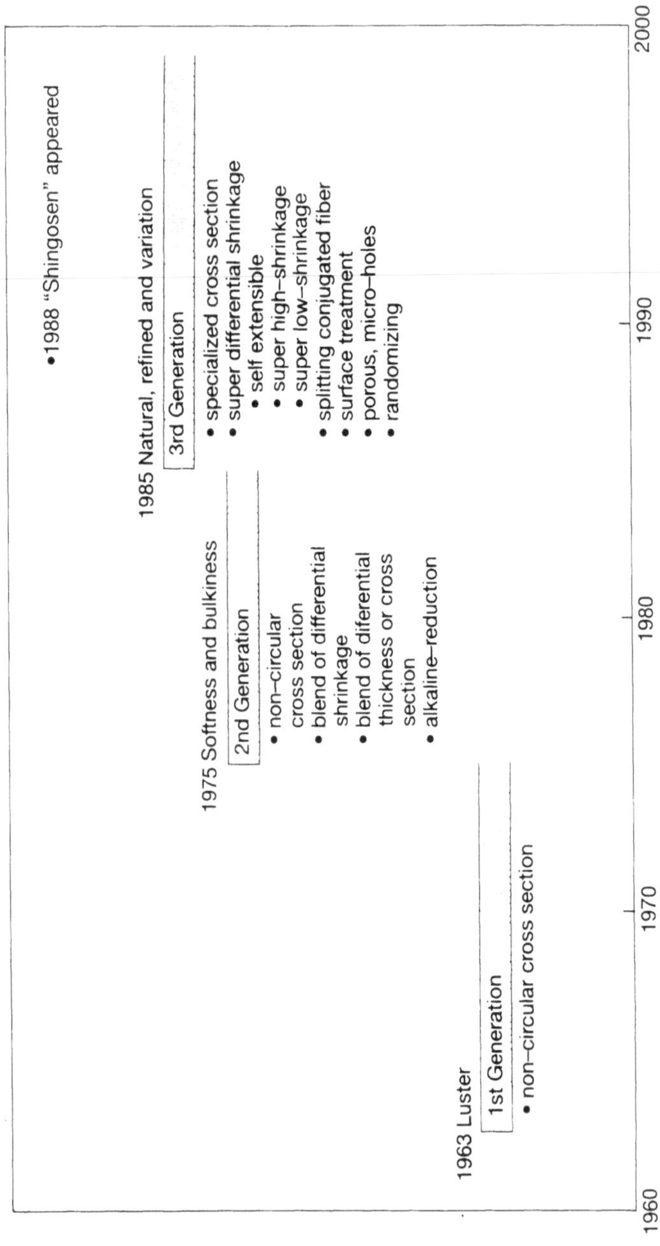

5.1 The evolution of silk-like fibers and fabrics.

In the 1970s, second generation silk-like fabrics appeared with a much refined appearance, softness and bulkiness. When fabrics composed of a blend of two types of filaments with different shrinkabilities are heated during the finishing process, a difference in filament length occurs and makes the fabric much bulkier and softer. The fabrics are also treated with an aqueous solution of sodium hydroxide to remove about 25%, by weight, of the polymer. This alkaline reduction treatment gives the fabric excellent softness. These second generation silk-like fabrics became the most important material for women's blouses and dresses until 1990 because of the excellent appearance, handle and texture obtained by applying these technologies.

The third generation appeared in the 1980s through applying:

1 Highly specialized non-circular cross-sections.
2 Blends of filaments having 'super differential shrinkability'.
3 Splitting of bicomponent composite filaments.
4 Surface treatment to produce complicated porous or micro-groove structures in the surface of the filament.

New generation textiles of synthetic fiber called 'Shingosen' appeared as the third generation in 1988, and the 'Shingosen boom' generated in Japan spread into the rest of the world.

These textiles are produced through a series of processes, including polymerization, spinning, drawing, fiber-blending, texturing, weaving, knitting, dyeing and finishing. In this chapter, the technologies for fiber production will be described and, in addition, the combinations with texturing, dyeing and finishing technology will be explained.

Table 5.1 compares the major technologies employed for the Shingosen fabrics with those of the previous generation of silk-like fabrics.

5.2 Specialized non-circular cross-section

Regular melt-spun fibers have circular sections due to the round orifices of the spinneret. Most highly-aesthetic fibers have non-circular sections produced with non-circular orifices. Silk has random triangular sections as shown in Fig. 5.2, which gives the fabric an excellent luster and superior handle. Figure 5.3 shows an example of a section of a second-generation silk-like fabric in which two types of fiber having different thickness are blended. The triangular section shown in Fig. 5.3 is most widely used for silk-like fabrics.

Various types of specialized section have been developed to give various and delicate luster, handle and texture to the fabrics. Typical examples are shown in Table 5.2 and Fig. 5.4–5.10. These specialized sections are produced with special orifices, such as a slit, L-shape,

Table 5.1 Previous technologies and new technologies for the Shingosen fabrics

Process	Previous technologies for first and second generations	Newly developed technologies for the third generation: 'The Shingosen'
General	Single technology or its simple combination	Combination of the plural technologies to create highly refined products
Polymerization	Bright (non-pigment) Dull (pigments) Cationic dyeable polymers (Sulphonate group)	High-shrinkable copolymers, high-gravity polymers, unevenly-degradable polymers Easily-degradable polymers
Spinning and drawing	Triangular cross-sections. Thin fibers	Specialized, uneven and randomized cross-sections Self-extensible fibers, super-low shrinkable fibers, high-shrinkable fibers, specialized conjugated fibers, super-fine fibers
Blending filaments	Simple combination of heat treated and untreated filaments	Combination of super-high-shrinkable filaments, super-less-shrinkable filaments and self-extensible filaments
Texturing	Specialized false-twisting	Composite (mixed) false twisting, sheath–core or double-layer false twisting
Dyeing and finishing	Alkaline-reduction	Splitting conjugated fibers Heat treatment of high-shrinkable fibers, less-shrinkable fibers and self-extensible fibers Mild pile raising, surface treatment of fiber, randomizing

5.2 Cross-section of silk (SEM).

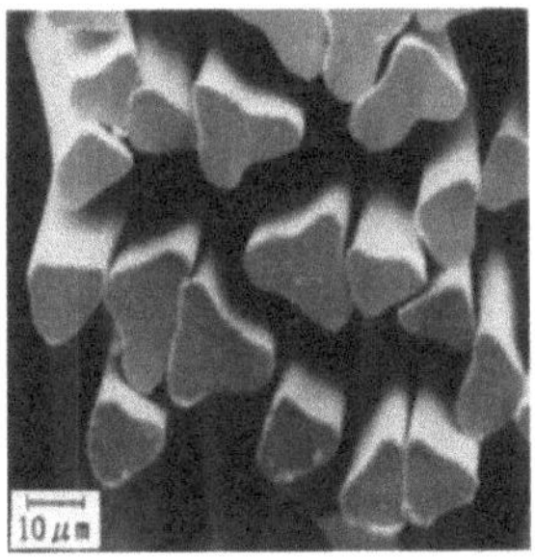

5.3 Triangular blended filaments of differential thickness, cross-section (SEM).

T-shape, H, W, X or similar shape modified slits, combinations of circle and slit, and arranged plural circles. Reference 2 (for specialized cross-sections) and Reference 3 (for conjugate or composite fibers) give further information.

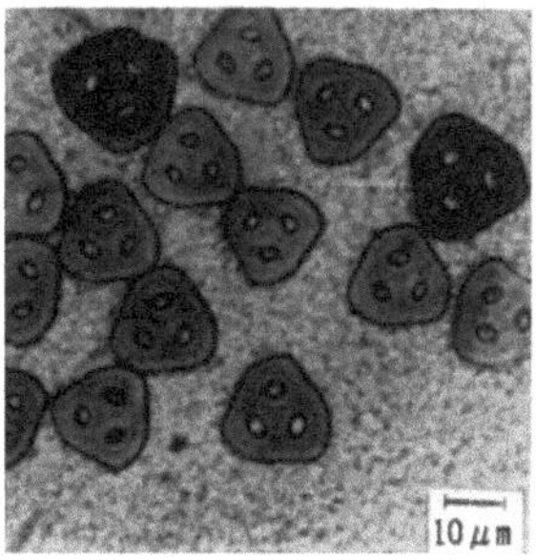

5.4 Hollow and triangular filaments, cross-section (optical microscope). Solo Sowaie by Asahi Chemical Industries Ltd.

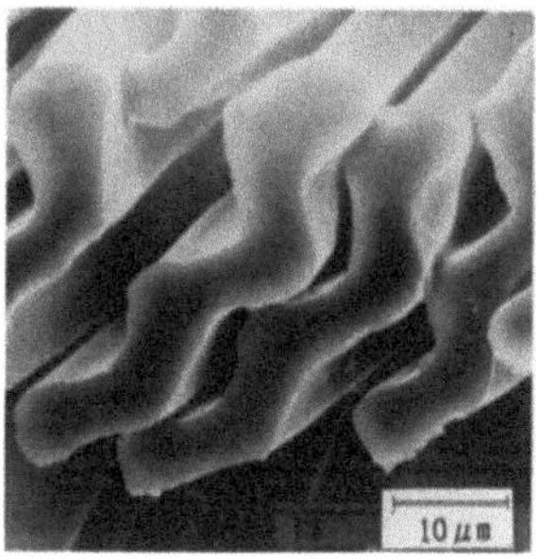

5.5 W-shaped flat filaments, cross-section (SEM). Fontana by Asahi Chemical Industries Ltd.

Table 5.2 Examples of specialized cross-section

Trade mark	Producer	Cross-section	Basic technology	Specialities of the product (Figure shows the cross section)
Solo Sowaie	Asahi Chemical Industries Ltd		Hollow, triangular, thick and thin	Higher bending stiffness, mild color (Fig. 5.4)
Fontana μ	Asahi Chemical Industries Ltd		W-shaped, self-crimping	Bulky, crispy, dry and cool hand (Fig. 5.5)
Soielise N	Kanebo Ltd		Pentagonal cross-section	Mild luster, dry hand, water-absorbent (Fig. 5.6)
Vivan	Kanebo Ltd		U-shaped cross-section, thick and thin	Mild luster, dry, spun-yarn-like, higher bending stiffness (Fig. 5.7)
Deforl	Kuraray Ltd		Flat cross-section, self-crimping	Deep color, bulky, higher bending stiffness (Fig. 5.8)
MSC	Unitika Ltd	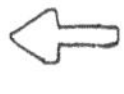	Arrow-like cross-section	Dry and cool hand (Fig. 5.9)
Mixy	Unitika Ltd		Random and multi-shaped cross-section	Dry hand, natural appearance, higher bending stiffness (5.10)

5.6 Pentagonal filaments, cross-section (SEM). Soielise N by Kanebo Ltd.

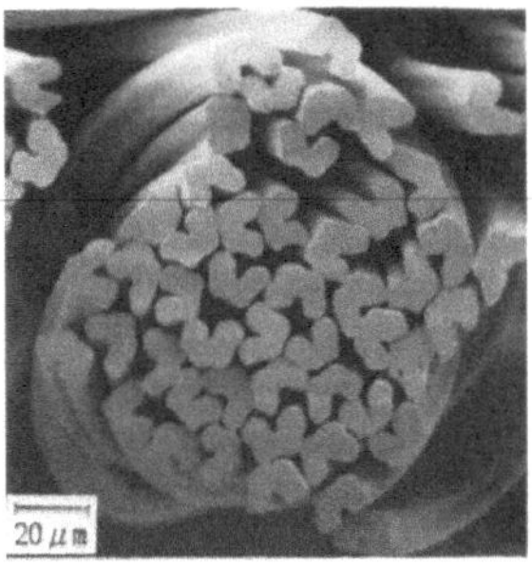

5.7 U-shaped filaments cross-section (SEM). Vivan by Kanebo Ltd.

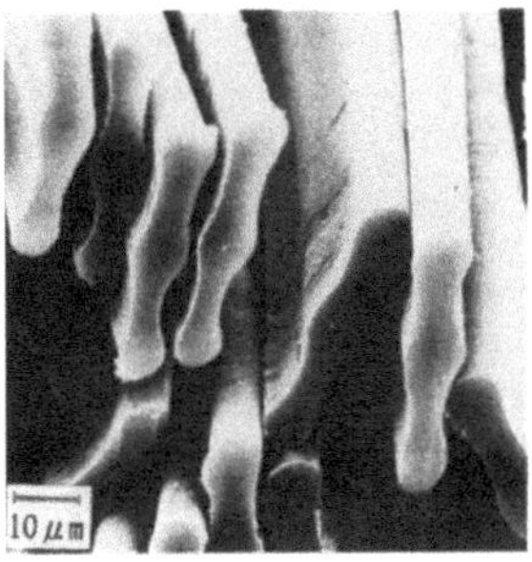

5.8 Flat filaments cross-section (SEM). Deforl by Kuraray Ltd.

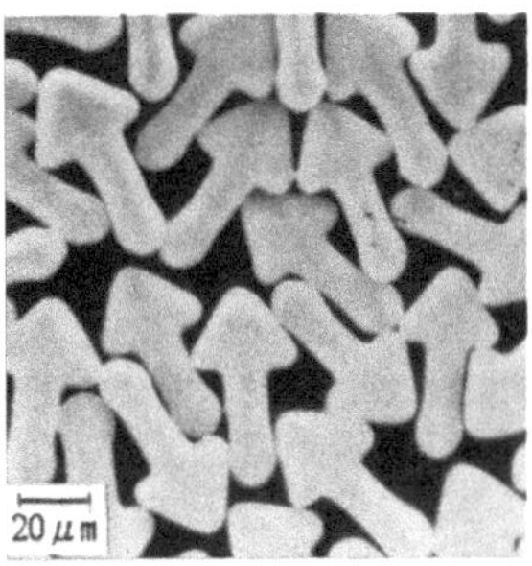

5.9 Arrow-shaped filaments cross-section (optical microscope). MSC by Unitika Ltd.

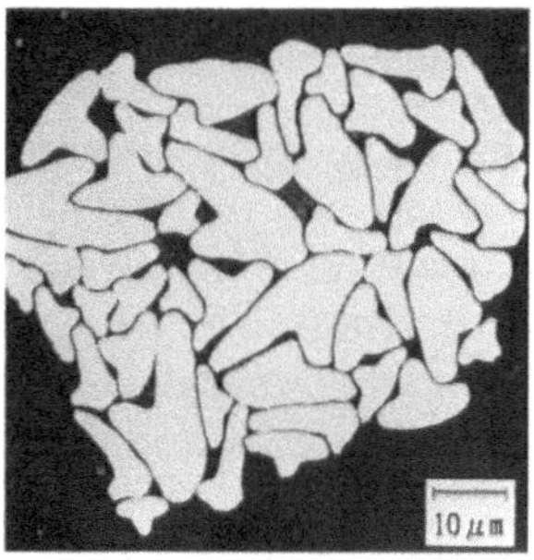

5.10 Random and multi-shaped filaments cross-section (optical microscope). Mixy by Unitika Ltd.

5.3 Blended yarns of different shrinkage

Woven fabrics composed of blended yarns which have 'differential shrinkage' are widely used for the second and third generation silk-like fabrics. Highly bulky and soft structures can be obtained, as shown in Fig. 5.11, by heating blended yarns in which highly shrinkable filaments and less shrinkable filaments are mixed together.

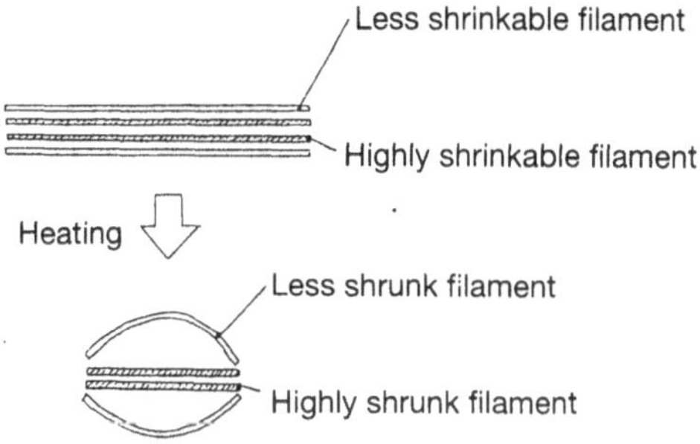

5.11 Heat-shrinking of blended yarn of high-shrinkage and low-shrinkage filaments.

The blending of a number of filaments may be carried out during the spinning, drawing, doubling, twisting or false-twisting processes. Various kinds of air-jet nozzle are also employed to produce uniformly or randomly blended and entangled filaments which give the finished fabrics various and delicate appearances and textures.

Highly shrinkable filaments are produced by copolymerization which reduces the crystallinity of the polymer, and/or by removal of the heat treatment process after drawing of the filaments. Less shrinkable filaments are, on the other hand, produced by stronger heat treatment after the drawing, or by 'super-high-speed spinning' which has been developed recently.

Most conventional differential-shrinkage yarn is a mixture of filaments heat treated after the drawing and untreated filaments. Recently, super-high-shrink filaments made of a copolymer and self-extensible filaments have come into practical use. The differential-shrinkage yarn has been replaced by 'super-differential-shrinkage yarn' which gives the fabrics extremely high bulkiness, softness and excellent texture.

The principle of a self-extensible filament was known many years ago,[4] however, it was not applied until the appearance of the third generation silk-like fabrics. Polyester filament yarn is drawn at a lower temperature and with a smaller draw ratio than usual. These drawn filaments are relaxed (shrunk) at a low temperature (around 100°C) to produce a special filament yarn having a low crystallinity and orientation. This is the self-extensible filament yarn.

The self-extensible filaments extend themselves (become longer) due to an increase in crystallinity and orientation on treatment at a high temperature (more than 100°C) during the dyeing and finishing of the fabrics. Filament yarns that have various shrinkabilities and extensibilities are produced by controlling the crystallinity and orientation using the technologies of copolymerization, spinning, drawing and heat treatment. Typical properties of the combination of filaments of different shrinkages are shown in Table 5.3 and examples of products in Fig. 5.12–5.14.

Table 5.3 Examples of combination for differential shrinkage

Filament	Technology	Shrinkage in boiling water	Combination and difference of shrinkage	
Regular	Drawing, heat treated	8%	7%	Conventional blended yarn of differential shrinkage
High shrinkable	Drawing, non-heat treated	15%		
Super-high-shrinkable	Copolymer, non-heat treated	30%	22%	New blended yarn of super-differential-shrinkage
Less shrinkable	Super-high-speed spinning	2%	13% 25%	
Self-extensible	Low drawing ratio, mild heat treatment and relaxing	−10%	40%	

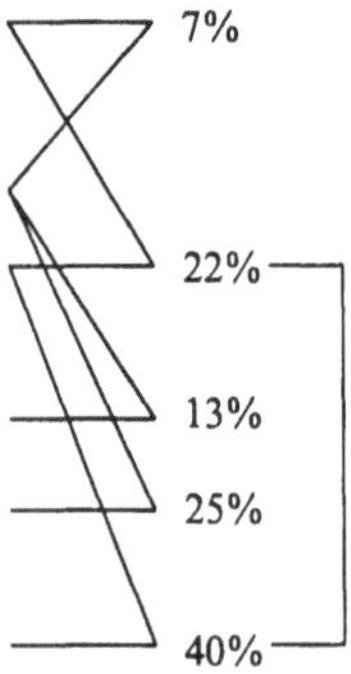

5.12 Fabric made of super differential shrinkage yarn (SEM). Legere by Kanebo Ltd.

5.13 Fabric made of self-extensible filament yarn (SEM). Ajenty by Teijin Ltd.

5.14 Fabric made of multi-step shrinkable filament yarn (SEM). Sillook Sildew by Toray Industries Inc.

5.4 Mixed-spinning, conjugate spinning and surface treatment

Inorganic particles or organic materials (mainly polymers) are blended into the polymer in the mixed-spinning process. One of the objectives of mixed-spinning is to increase the specific gravity of the fiber and improve the drapeability of fabrics by mixing with inorganic particles which have a high specific gravity. Metallic compounds having high specific gravity and white colour, such as titanium oxide, zinc oxide and barium suphate, are suitable. A specific gravity of 1.5 is the most desirable from the viewpoint of drapeability. Generally, the mixing is carried out at the polymerization process, but it can also be done between polymerization and spinning, or at the spinning process.

The content of inorganic particles should be less than 10% by weight, usually less than 5% to avoid the risk of frictional damage to the apparatus during the spinning, drawing, weaving or knitting processes due to the inorganic particles of the fiber. A particle diameter of less than 1μm, and preferably less than 0.5 μm, is desirable to obtain better spinnability.

Most fabrics of high specific gravity have a dull luster due to the mixed pigment particles. The surface of alkaline-treated fabrics of high specific gravity fibers has innumerable micro-holes due to the higher degradation rate of the polymer near the particles. The porous structure gives the fabric a dry and cool handle. This type of Shingosen is called 'Rayon type' because of its high drapeability, dull luster and cool and dry feeling. Examples of surface treatments are shown in Fig. 5.15–5.17.

The other purpose of mixed-spinning is to produce numerous micro-holes, micro-craters and/or micro-grooves by a post-treatment (alkaline reduction), giving the finished fabrics a deep colour, a good appearance, and a good handle. Various kinds of metallic compounds such as titanium oxide, zinc oxide, kaolin, alumina, calcium carbonate, barium sulphate, zeolite and silica can be applied for this purpose. Various kinds of polymers having different alkaline degradation rates are also mixed-

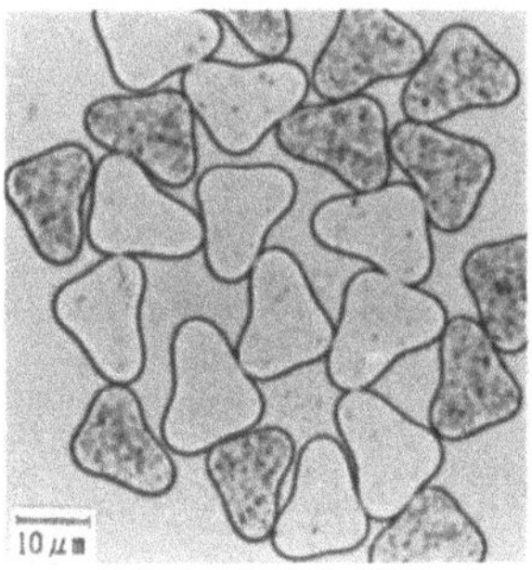

5.15 Blended filaments of differential shrinkage and gravity. Sowaie Pairny by Asahi Chemical Industries Ltd.

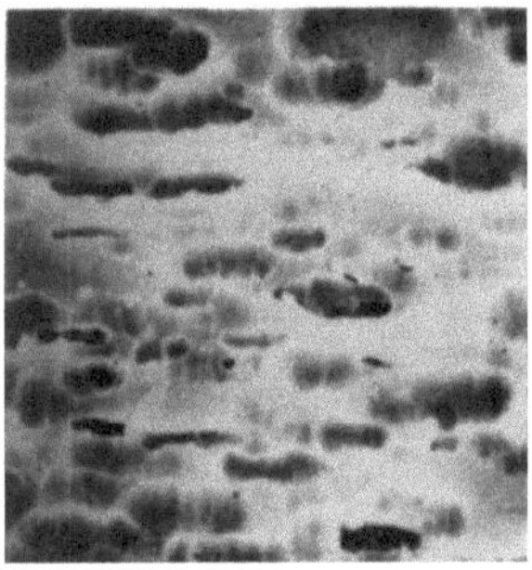

5.16 Surface of high-gravity fiber having micro-craters (SEM). XY-E by Kuraray Ltd.

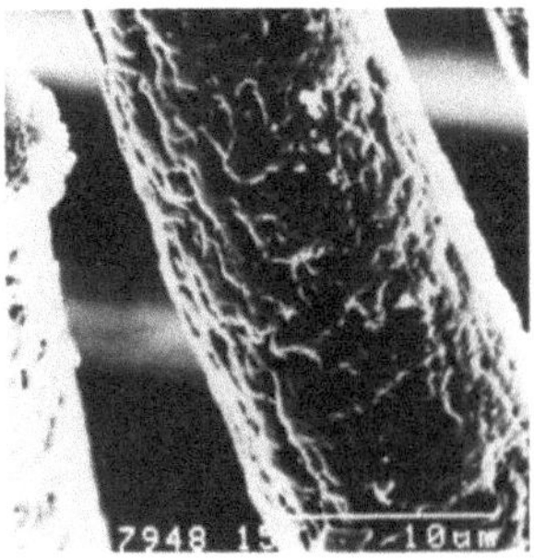

5.17 Surface of high-gravity fiber after treatment (SEM). Louvro by Toyobo Ltd.

spun or conjugate-spun to form microporous structures or micro-grooves in the surface of the fiber. Examples of combinations of mixed-spinning, conjugate-spinning and surface treatments, as described in Table 5.4, are shown in Fig. 5.18–5.23. Reference 5 gives further information about the mixed-spinning of polymers.

Table 5.4 Examples of the products of conjugated or mixed spinning combined with surface treatment technology

Trade mark	Producer	Fiber	Cross-section	Cross-section, surface after finishing	Specialty of products
Treview	Kanebo Ltd	Random conjugated spinning		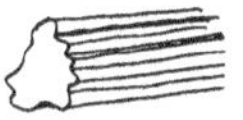	Dry, spun-silk-like natural feeling
Fontana	Asahi Chemical Industries Ltd	Mixed or conjugated spinning		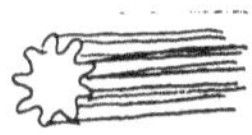	Dry-spun-silk-like natural feeling
SN 2000	Kuraray Ltd	Inorganic particle mixing	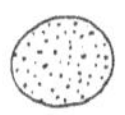		Dry hand, deep color
Sillook Royal	Toray Industries Inc	Radial conjugated spinning	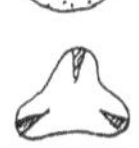	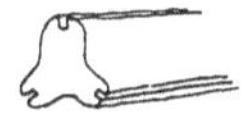	Bulkiness silk sound
Sillook chatelaine	Toray Industries Inc.	Inorganic particle mixing			Dry hand, rayon-like
Louvro	Toyobo Ltd	Inorganic particle mixing		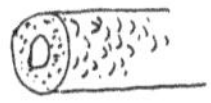	Dry hand, rayon-like, higher bending stiffness
Rapitus	Teijin Ltd	Copolymer or mixed spinning	?		Natural feeling, crispy, spun-silk-like

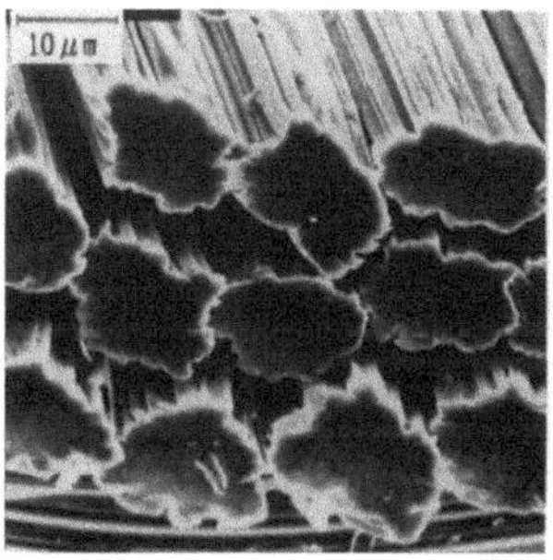

5.18 Randomized fiber having micro-grooves in the surface (SEM). Fontana by Asahi Chemical Industries Ltd.

5.19 Randomized fiber (SEM). Treview by Kanebo Ltd.

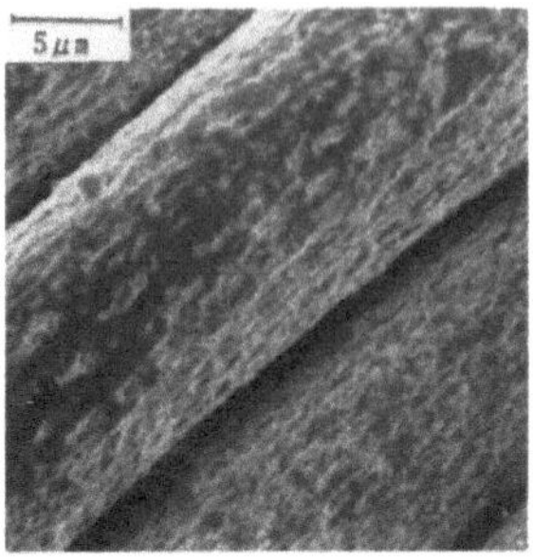

5.20 Deep color fiber having micro-craters in the surface (SEM). SN2000 by Kuraray Ltd.

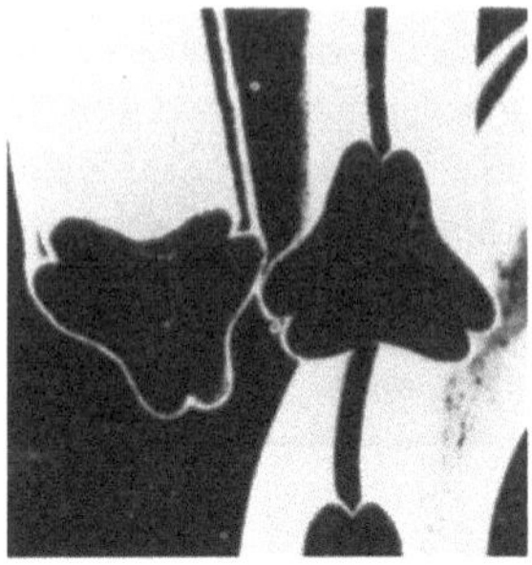

5.21 Triangular filaments with grooves (SEM). Sillook Royal by Toray Ind. Inc.

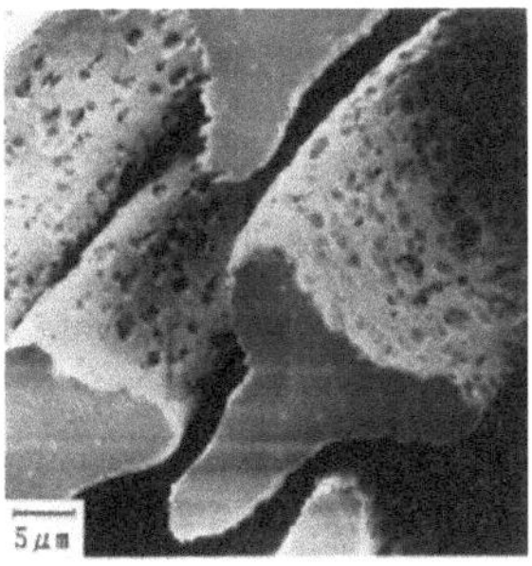

5.22 Triangular filaments having micro-holes in the surface (SEM). Sillook chatelaine by Toray Industries Inc.

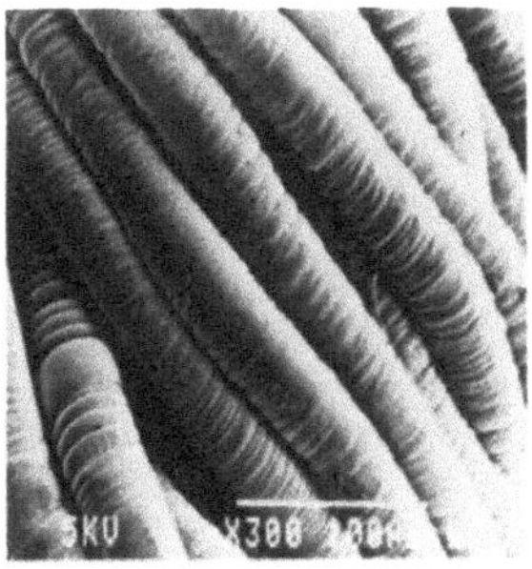

5.23 Filaments having micro-scales in the surface (SEM). Rapitus by Teijin Ltd.

5.5 Splitting of conjugate fibers and super-fine fibers

One of the most important applications of conjugate (composite) fibers is the production of super-fine fibers and special cross-sections by the splitting of the original conjugate fibers. The five major applications of super-fine fibers are:

1 Artificial suedes.
2 Second generation artificial leathers.
3 Moisture-permeable, water-repellent, high density fabrics.
4 Silk-like fabrics (materials for dresses and blouses).
5 High-performance wiping cloths.

Reference 6 gives further information about wiping cloths; see also Chapter 9. Some examples of types 3 and 4 for fashionable garments will be discussed here.

Bicomponent conjugate fibers are split by:

1 Dissolution or degradation of a component polymer.
2 Separation of the two components caused by swelling and shrinkage of a component.
3 Separation of the components by mechanical distortion.

Therefore, combinations of two components which have a large difference in solubility, degradability or swelling and combinations which have poor adhesion are chosen for the conjugate fibers. Furthermore, a suitable conjugate arrangement must be chosen to promote easy separation.

Figure 5.24 shows a typical, radially conjugated fiber being split into eight triangular segments (polyester) and a radial segment (polyamide 6). Figure 5.25 shows a fabric made from the radially conjugated fiber. It has an extremely soft and smooth handle and delicate appearance due to a super-fine pile (micro pile). In this case, the component shrinks only a little. This causes the polyester super-fine fiber to be raised from the substratum structure to form a pile, making the fabric very bulky and soft.

Figure 5.26 shows a flower-like conjugate fiber in which two components are combined. One of the components, for the core and the eight petal-like segments, is polyethylene terephthalate. The other component, between the polyester segments, is a readily soluble or degradable modified polyester. The readily soluble polyester has a very high rate of degradation, ten to a hundred times that of regular polyethylene terephthalate. Figure 5.27 shows a fabric in which highly shrinkable filaments are blended so as to raise the super-fine fibers made

5.24 Radially conjugated filaments being split (SEM). Belma-X by Kanebo Ltd.

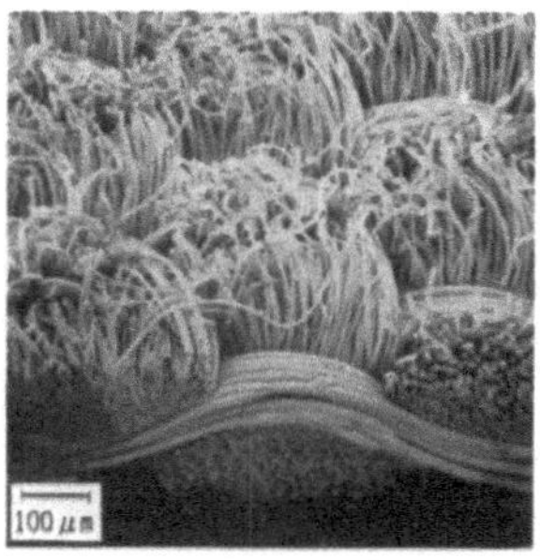

5.25 Super-high-density water-repellent fabric having micro-pile (SEM). Belseta PS by Kanebo Ltd.

5.26 Flower-like conjugate filaments (SEM). Cosma-a by Kanebo Ltd.

5.27 Super-fine fabric for ladies' blouses (SEM). Nazca by Kanebo Ltd.

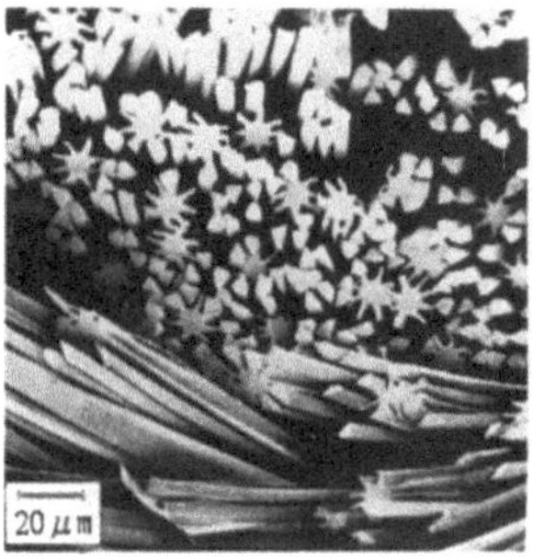

5.28 Cross-section of super-fine fabric (SEM). Piceme by Toray Industries Inc.

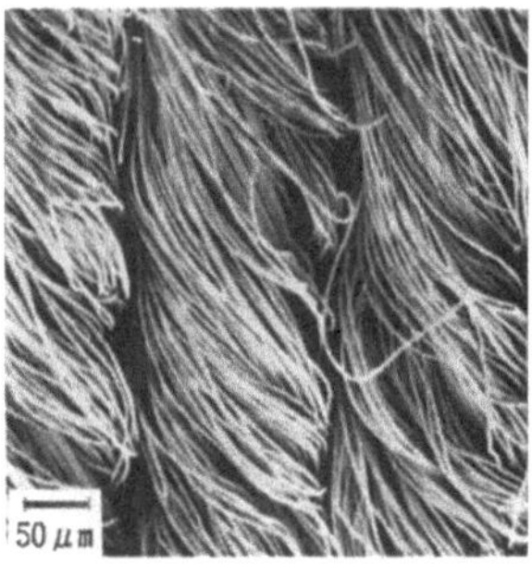

5.29 Surface of super-fine fiber fabric (SEM). Rominar by Unitika Ltd.

by the splitting of the flower-like conjugate fiber to form a pile, giving the structure higher softness and bulkiness.

Figure 5.28 shows the cross-section of a woven fabric produced by the splitting of another type of radially conjugate fiber. Radial segments and triangular super-fine filaments can be observed in the picture.

Figure 5.29 shows the surface of a fabric composed of super-fine filaments (on the surface) and highly shrinkable filaments (inside). Some super-fine fiber is produced by direct melt spinning, which is limited to a linear density of 0.4 denier. Super-fine filament yarn having a linear density of 0.1 denier, for example, can be produced today by highly advanced melt spinning technologies. However, it is very difficult to handle the super-fine filament yarn at the drawing, texturing, knitting and weaving process. Therefore most knitted or woven fabrics made from super-fine fibers are produced by splitting the conjugate fibers after the knitting or weaving process.

5.6 Funny fibers

Recently, 'funny fibers' or 'very interesting fibers' became topics for the fiber industry. The funny fibers or fabrics have practical, or sometimes impractical, properties such as thermochromic, photochromic, perfumed, antibacterial, deodorant, heat-storing, water-repellent, water-absorbent and electro conductive effects.

Some of these funny fibers have recently been developed by specialized spinning technology, such as perfumed conjugate fibers composed of a terpene-containing polymer and a fiber-forming polymer, anti-bacterial fibers in which zeolite particles containing silver ions are blended, light-to-heat transferring fibers in which carbonised zirconium particles are blended, and water absorbent fibers having a hollow and porous structure or a slit as a water passage. Reference 7 gives further information.

References

1 Kawabata S, *Analysis and Standardisation of Hand Evaluation*, (2nd Edition), The Textile Machinery Society of Japan, 1980.
2 The Society of Fibre Science and Technology, Japan, *Illustrated Shapes and Structures of Fibre*, Asakura-Shoten, Tokyo, 1982.
3 Matsui M, *Sen-i Kikai Gakkaishi*, **34**, 319, 1981.
4 Jap. Pat. (Examined), 66-12,052.
5 The Society of Polymer Science, Japan, *High Performance Polymer Alloys*, pp 265–93, 1991.
6 Matsui M, *Kako Gijutsu*, **24**, No. 2-3, 1989.
7 Matsui M, *Kako Gijutsu*, **22**, No. 1-2, 1987.

6

Fiber spinning of anisotropic polymers

H H Yang and S R Allen
Du Pont, Richmond, USA

6.1 Anisotropic polymers

A key factor in the preparation of high strength fibers is the formation of anisotropic solutions or melts from anisotropic polymers. Many aromatic polyamides, polyhydrazides, polyesters, polyazomethines, polyimides and heterocyclic polymers with extended chain conformation form anisotropic solutions and melts. These solutions and melts exhibit liquid crystal behavior at adequate solution concentrations and temperatures. During fiber formation, liquid crystal domains in these polymer solutions or melts will readily undergo orientation under the influence of shear and elongational flow. Anisotropic polymers also undergo phase transition during cooling to form a highly crystalline solid. Such behavior results in a fiber of highly oriented, highly crystalline structure and high strength. For this reason, solution and melt anisotropy is a desirable but not necessary polymer property in the formation of high-strength fibers.

6.1.1 Classification of anisotropic polymers

There are two types of anisotropic polymers: ‘lyotropic’ and ‘thermotropic’ polymers. Lyotropic polymers are those whose solutions respond to solvent type and concentration changes. They are represented by many aromatic polyamides, polyhydrazides, polyimides, and heterocyclic polymers containing chain-extending repeat units. Thermotropic polymers are those whose melts behave like liquid crystals in response to temperature changes. They are represented almost exclusively by many aromatic polyesters containing mesogenic repeat units. Thermotropic polymers are often referred to as liquid crystalline polymers (LCP). We will use the generic term ‘anisotropic polymers’ in this chapter to refer to both lyotropic and thermotropic polymers. Many investigators have reported on lyotropic and thermotropic polymers in recent years.[1–18]

6.1.2 Chemical structures

Anisotropic polymers exhibit mesomorphic behavior in solutions or melts like low molecular weight liquid crystals. The difference is that anisotropic polymers have much higher molecular weights. In 1965, Kwolek[19] and other scientists[20–25] at the Du Pont Company first observed the liquid crystalline behavior of extended-chain aromatic polyamides in anisotropic solutions. Many anisotropic polyamides were prepared and spun into fibers. These polymers contained a variety of stiff chain structures but the common feature was that the components exhibited essentially coaxial or parallel extension of the chain forming bonds. Of these, poly (*p*-benzamide) (PBA) and poly (*p*-phenylene terephthalamide) (PPD-T) are the best known lyotropic polymers:

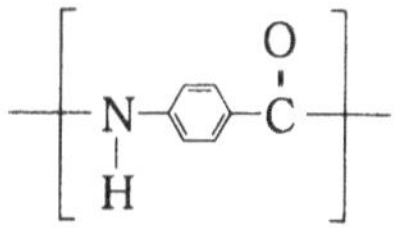

PBA

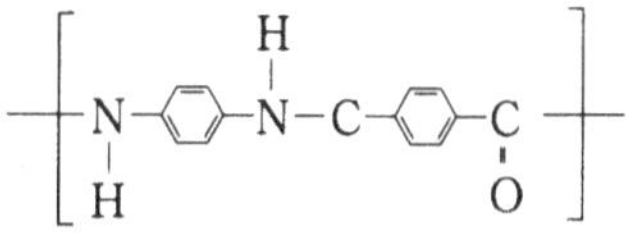

PPD-T

That discovery led to the commercialization of poly (*p*-phenylene terephthalamide) fiber under the tradename of KEVLAR in 1972. In the ensuing two decades, scientists at Carborundum, Eastman Kodak, Celanese and Du Pont discovered many aromatic polyesters with thermotropic melt behavior. These discoveries generated worldwide interest, which has led to many excellent high performance fibers and thermoplastics from thermotropic wholly aromatic polyesters. There are now several classes of aromatic polymers which exhibit lyotropic or thermotropic behavior. They include

Aromatic polyamides
Aromatic polyhydrazides
Aromatic polyesters
Aromatic polyazomethines
Aromatic polyimides
Aromatic heterocyclic polymers

Table 6.1 presents examples of these anisotropic polymers. It is interesting to note that anisotropic aromatic polyamides, polyahydrazides, polyimides and heterocyclic polymers are inherently high melting and are commonly lyotropic. Wholly aromatic polyesters generally melt at 300–350°C and are commonly thermotropic, as are polyazomethines.

The ability of a polymer to form liquid crystals is closely related to its structural features. There are therefore some preferences on polymer

Table 6.1 Anisotropic polymers and fibers[18]

Polymer	Fiber type	Density g/cc	Ten. g/d	Mod. g/d	T_g °C	T_m °C	T_d °C
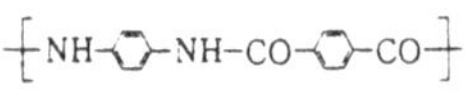	KEVLAR 29	1.43	23	580	> 375	550	555
	KEVLAR 49	1.45	23	950			
	KEVLAR 119	1.44	24	470			
Aromatic polyamide	KEVLAR 129	1.45	26.5	750			
	KEVLAR 149	1.47	18	1100			
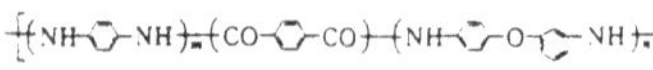 Aromatic copolyamide	Technora	1.39	28	590		500	
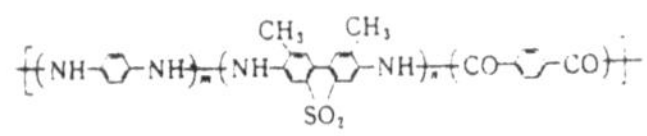	As spun		24	490			455
Aromatic copolyamide	Heat treated		31	875			
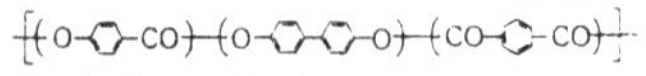 Aromatic copolyester	Ekonol	1.4	31	1100		–350	
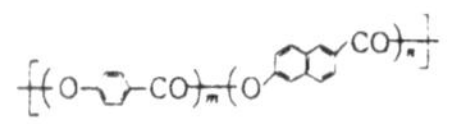 Aromatic copolyester	Vectran	1.47	22–25	600–700		–250	
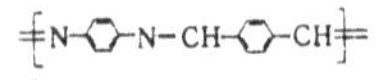 Aromatic polyazomethine		1.39	38	1012		250	
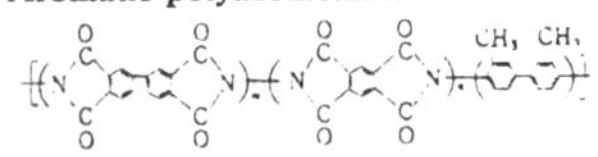 Aromatic copolyimide		1.34	27	1659			
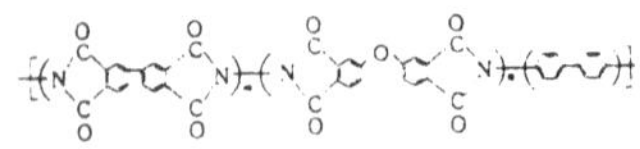 Aromatic copolyimide		1.47	22	1600			
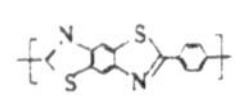 Aromatic heterocyclic polymer		1.57	25	2690			600
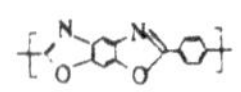 Aromatic heterocyclic polymer		1.57	25	2900			650

compositions and modifications to ensure liquid crystallinity. Generally, rigid chain polymers with linear chain conformation and high length/diameter ratio are favored. Figure 6.1 illustrates the rod-like, nearly coplanar chain conformation of poly (*p*-phenylene terephthalamide). However, a polymer may have too low solubility or too high melt temperature to be anisotropic. In that case, it must be modified structurally to enhance its solubility or meltability. It is very important

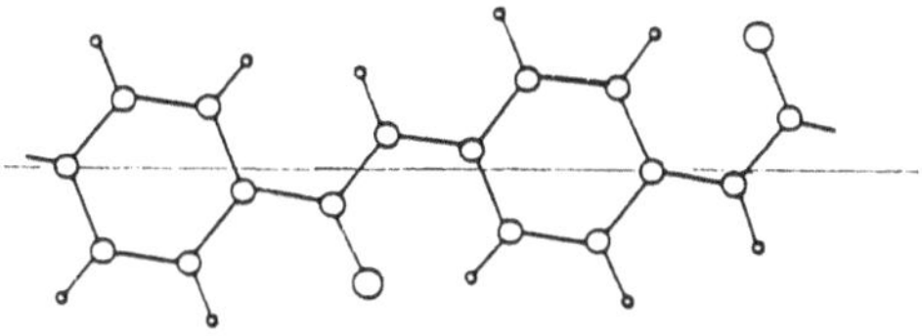

6.1 Molecular model of poly (*p*-phenylene terephthalamide).

to maintain a good balance between such structural modification and linear chain conformation in order to retain sufficient anisotropy.

6.1.3 Physical behavior

Liquid crystals of low molecular weight compounds exist in three major phases: nematic, smectic, and cholesteric. These phases are shown schematically in Figure 6.2.[8] The nematic state, which denotes a 'thread' pattern, contains domains of well-ordered macromolecules. The nematic ordered structure generally exhibits an optical texture of threads when it is viewed in a polarized light field. The smectic state, which denotes a layer of 'soap', contains a layered structure of ordered macromolecules.

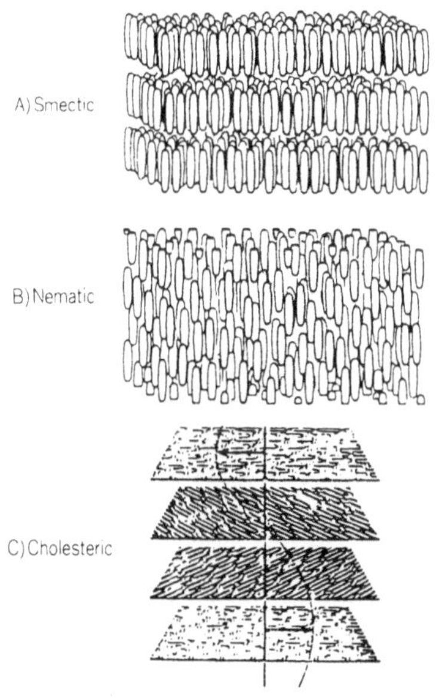

6.2 Phase structure of liquid crystalline polymers.[8]

It is further subdivided according to minor differences in optical features. Sackman and Demus[26] identified smectic subclasses by the order of discovery, i.e., Sa, Sb, ..., Sh. De Vries[27] classified them as α, β, γ, ..., according to the order of increasing molecular packing. The cholesteric state is a subclass of the nematic state. It contains layers of nematic molecules where each layer is rotated with respect to its neighboring layers, giving rise to an overall helical structure.

The behavior of an anisotropic polymer depends on a number of variables: temperature, solution concentration, solvent, and molecular structure. A thermotropic system will undergo simple phase transitions during heating from its crystalline solid state to the smectic state, the nematic state, and finally the isotropic state unless decomposition occurs. These phase transitions are reversible:

$$\text{Crystalline solid} \rightleftarrows \text{Smectic} \rightleftarrows \text{Nematic} \rightleftarrows \text{Isotropic}$$

at increasing temperature. Since the liquid crystal state lies between the phase boundaries of solid crystals and isotropic liquids, it is often referred to as the 'mesomorphic' structure or a 'mesophase' to describe its intermediate form.

As with low molecular weight liquid crystals, a lyotropic polymer yields an anisotropic solution when its solution concentration exceeds a critical value. In general, the critical concentration decreases with increasing polymer molecular weight. The anisotropic solution exhibits the isotropic-mesophase transition upon cooling. Its anisotropic characteristics generally include optical depolarization, reduced solution viscosity, increased density, and detectable crystalline-like transition temperatures.

Similarly, the isotropic melt of a thermotropic polymer will become anisotropic when it is cooled to a certain temperature. The anisotropic melt is also characterized by optical depolarization, reductions in melt viscosity and melt temperature, and well-defined crystalline-like phase transitions.

Like conventional polymers of polydisperse molecular weight distribution, most anisotropic solutions and melts undergo subcooling during freezing. They pass through a liquid–solid biphasic region before they are completely solidified. For this reason, the onset of liquid–solid transition is often considered as the melt temperature, and the completion of liquid–solid transition is taken to be the freezing temperature. The isotropic–mesophase temperatures of both lyotropic and thermotropic polymers often differ from their liquid–solid transition temperatures. Some lyotropic solutions are nematic in the liquid phase; others are in the solid phase. Most thermotropic polymers exhibit the melt mesophase above their melt temperatures.

The rheological and orientation behavior of liquid crystalline polymers appears to be mostly attributable to macromolecular structures. Consequently, a domain theory is often invoked to describe randomly oriented domains containing highly ordered liquid crystals in a less ordered matrix. Figure 6.3[10,18] illustrates schematically such a concept. Most domains are considered to be of an elongated ellipsoid shape. Some are thought to be disc shaped. The domain sizes are generally accepted as in the order of microns.

6.1.4 Quantitative characterization

There are several physical parameters which can be used to characterize anisotropic systems. Although these parameters are highly theoretical, they are very useful in describing the anisotropy of a polymer system, at least conceptually.

6.1.4.1 Order parameter

The level of local structural order in an anisotropic system is defined by an order parameter, S, according to the following equation:

$$S = \frac{3\cos^2\theta - 1}{2} \qquad [6.1]$$

where θ is the angle between the long axis of a molecule and the axis of preferred orientation. The value of S is zero for isotropic systems and is 1.0 for uniaxial orientation. Low molecular weight polymers generally have an S value of 0.4–0.7 which corresponds to a root mean square orientation angle of 39–27°. For anisotropic polymers, particularly thermotropic polymers, the order parameter S can be as high as 0.85, but is highly affected by many variables such as temperature and composition.

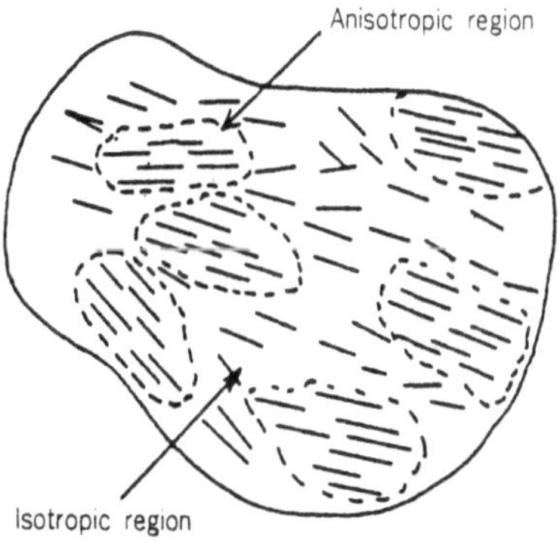

6.3 Schematic representation of domain theory.[10,18]

6.1.4.2 Flory theory

The effect of concentration on solution anisotropy of a lyotropic polymer can be interpreted in terms of molecular packing in a given volume. When the solution concentration is increased, polymer molecules in the solution become increasingly packed together. This is accompanied by an increase in the intermolecular van der Waals and bond forces. For rigid chain polymers with a high degree of molecular asymmetry, it eventually leads to the formation of liquid crystal structure. Flory[28] postulated in 1956 that a solution will become anisotropic when it reaches a critical volume concentration, ψ^*, according to the following equation:

$$\psi^* = \frac{8}{x}(1 - \frac{2}{x}) \qquad [6.2]$$

where x is the axial ratio of molecular conformation. This is the famous Flory equation which predicts that ψ^* will decrease with increasing axial ratio x or molecular rigidity. Thus, a polymer with a high axial ratio is preferred for the formation of anisotropic solution.

6.1.4.3 Persistence length

The persistence length is the end-to-end length of a polymer chain in a dilute solution and is a measurement of chain rigidity. It is defined by the Benoit equation.[29,30] The persistence lengths of anisotropic aromatic polymers have been studied by a number of investigators. Several values are presented in Table 6.2.[18] It is interesting to note that lyotropic polymers generally give greater persistence lengths than thermotropic polymers. This is consistent with the fact that lyotropic polymers have greater chain stiffness, higher glass transition and melt temperatures than thermotropic polymers.

6.1.4.4 Mark–Houwink equation

Except for a few well-known polymers, the rheological properties of liquid crystalline polymers have not been reported extensively. The

Table 6.2 Persistence lengths and Mark–Houwink exponents[18]

Polymer	Persistence length, ρ, Å	Mark–Houwink exponent, *a*
Flexible polymers		0.5
Aliphatic – aromatic polyesters	50 + 10	
Aromatic polyamides		
PBA	410	1.09
PPD–T	175–450	1.7
Thermotropic aromatic polyesters	170	0.76–0.73
HBA/HNA	60–90	0.98
Aromatic polyazomethines		
MePPD–TA	150 + 25	1.20

Mark–Houwink equation is often referred to as a general equation of viscosity:

$$[\eta] = KM^a \qquad [6.3]$$

where $[\eta]$ is the intrinsic viscosity, M is the molecular weight, K and a are both constants. The exponent a varies from 0.5, for flexible chain polymers, to 1.7 for rigid chain polymers.[31] Several a values are presented in Table 6.2.

6.1.4.5 Aromaticity parameter

Calundann and Jaffe[15] devised an aromaticity parameter to describe the degree of aromaticity of a polymer. It is defined as the ratio of hybridized carbons to the total number of atoms in the repeat unit. They showed the tensile moduli of aliphatic-aromatic polyesters increased linearly with increasing aromaticity. Such a correlation signifies that the polymer chain stiffness increases with increasing aromatic segments in the polymer main chain.

In addition to the above quantitative methods, several alternate methods have been used to characterize anisotropic polymers. They include wide angle x-ray diffraction, NMR spectroscopy, rheo-optical measurements, and others.

6.1.5 High strength, high modulus fibers

Fibers of anisotropic polymers are generally characterized by their high strength and high modulus. They have been found useful as high performance fibers for use in advanced materials. There is now a considerable amount of worldwide industrial activities on the development of high performance fibers from anisotropic polymers. Many kinds of synthetic fibers are competing for a share in the advanced materials market. While anisotropic fibers do not share in the advanced materials market, they do reflect a considerable degree of sophistication in polymer chemistry and spinning technology. The methods used to produce anisotropic high-strength fiber vary from polymer to polymer. In principle, the process of fiber preparation begins with polymer synthesis, followed by polymer isolation and solvent recovery, preparation of spin solution if necessary, and finally, fiber formation by spinning. These process steps are sometimes very complicated.

Black[32] noted a reciprocal relationship between the initial modulus, M, and elongation at break, E_b, of synthetic fibers. It can be expressed by the following equation:

$$M = 1000/0.68E_b \qquad [6.4]$$

As a whole, fibers of anisotropic polymers have considerably higher strength and modulus, but much lower elongation at break than conventional fibers. Most as-spun aromatic high-strength fibers have elongation of 3–6% and modulus of 150–500 g/d. Heat treated aromatic fibers have elongation of 1.5–4% and modulus of 600–2500 g/d. The high fiber strength and modulus reflect the high levels of crystalline orientation and crystallinity of anisotropic fibers. These microstructural features are greatly dependent on polymer compositions and fiber processes. Therefore, the modulus–elongation relationship of various classes of anisotropic high-strength fibers may be similar at relatively low modulus levels as described by equation 6.4, but quite different at high modulus levels.

Most high strength, high modulus fibers offer both high tensile strength and high tensile modulus. High fiber strength is often accompanied by high fiber modulus. Conversely, high fiber modulus does not always ensure high fiber strength. This is why one should preferably seek high fiber strength in the development of spinning technology.

6.2 Lyotropic polymers

6.2.1 Phase behavior

The phase behavior of lyotropic systems is complicated by the solvent type, solution concentration, polymer molecular weight, and temperature. Generally, the phase transition temperature of a lyotropic solution increases with increasing concentration and polymer molecular weight. The phase behavior is also complicated by the polymer structural conformation and intermolecular interactions. These parameters exert overlapping effects on the solubility, glass transition and melt temperatures of a polymer, which in turn affect the lyotropic phase transition behavior.

The nematic and cholesteric mesophases of lyotropic polymers can be readily identified by microscopic examination. However, the existence of a smectic mesophase is not well defined but only suggestive in some cases. For this reason, it is often difficult to determine the critical concentration or transition temperatures of a lyotropic polymer precisely. Some polymers even degrade below the nematic-isotropic transition temperature so that it is impossible to determine those transition temperatures.

A good example of lyotropic solution is that of poly (*p*-phenylene terephthalamide) (PPD-T) in sulfuric acid. Figure 6.4 shows the viscosity–concentration relationship of such a solution of PPD-T with a moderate molecular weight. At low polymer concentrations, the solution viscosity increases with increasing concentration like the isotropic solution of a flexible chain polymer. Above a critical concentration of 12%, the solution viscosity decreases abruptly with increasing concentration. This

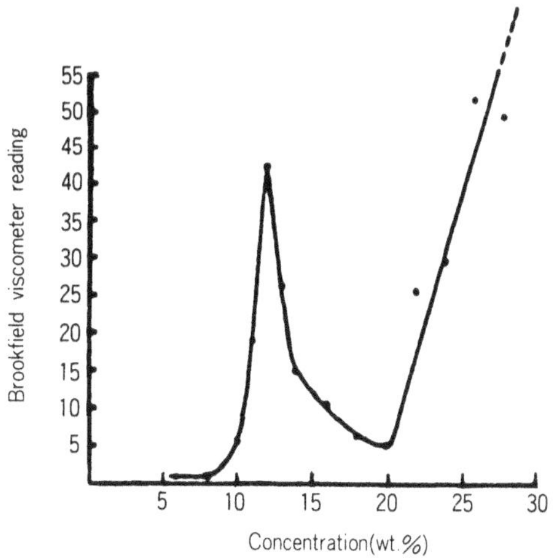

6.4 Bulk viscosity vs concentration of PPD-T/H_2SO_4 solution.[25]

behavior is caused by the close packing of polymer molecules and the formation of ordered domains at increasing concentrations. The solution viscosity reaches a minimum point at about 20% and then reverses to an uptrend at further concentration increase. A solid phase will ultimately appear in the solution when it is oversaturated.

The anisotropic PPD-T/H_2SO_4 solution exhibits liquid crystal behavior. It has the flow properties of a liquid, and is crystal-like with the ability to depolarize a cross-polarized light. When the solution is subjected to a shear or elongation flow, the liquid crystal domains will undergo further ordering in the flow direction and achieve a higher order of molecular orientation. This behavior is easily seen by optical microscopy. For illustration, Figure 6.5 presents an optical micrograph of PPD-T/H_2SO_4 solution at rest in a cross-polarized light field. The liquid crystal domains transmit the polarized light by depolarization and are seen as the bright regions. The dark regions contain those domains incapable of light transmission and the isotropic regions. Figure 6.6 shows the same PPD-T/H_2SO_4 solution after shear by moving the microscope cover glass slightly. It can be seen that the liquid crystal domains have elongated somewhat and aligned in the direction of shear. Under high shear or elongation flow, the boundaries of liquid crystal domains may be partly destroyed and domains may link up to form a fibrillar structure when coagulated.

For fiber preparation, a lyotropic solution like that of Figure 6.4 should best be processed near the minimum solution viscosity and at the temperature closest to its nematic–smectic transition temperature. These conditions would enhance maximum solution ordering before fiber spinning.

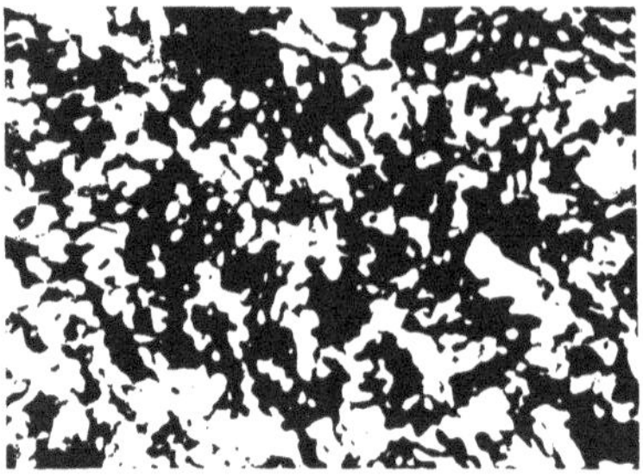

6.5 PPD-T/H_2SO_4 solution at rest.

6.6 PPD-T/H_2SO_4 solution after shear.

6.2.2 Rheological properties

Information on the rheological properties of a lyotropic solution is needed for the design consideration of spinneret and capillary holes, and for the fiber spinning process. Unfortunately, such information is rather scarce in the literature. The PPD-T/H_2SO_4 solution is perhaps the only system that has been studied extensively. However, studies on this system are largely limited to solutions less concentrated than the spinning solutions.

Lyotropic solutions generally exhibit viscoelastic behavior. They are non-Newtonian, and are pseudoplastic with shear thinning at increasing shear rate. For polymers of near-linear chain conformation, their lyotropic solutions are known to give less die swell and are less tractable than isotropic solutions.

The rheological properties of PPD-T/H_2SO_4 solutions have been studied in several investigations.[33–38] Figures 6.7 and 6.8 show

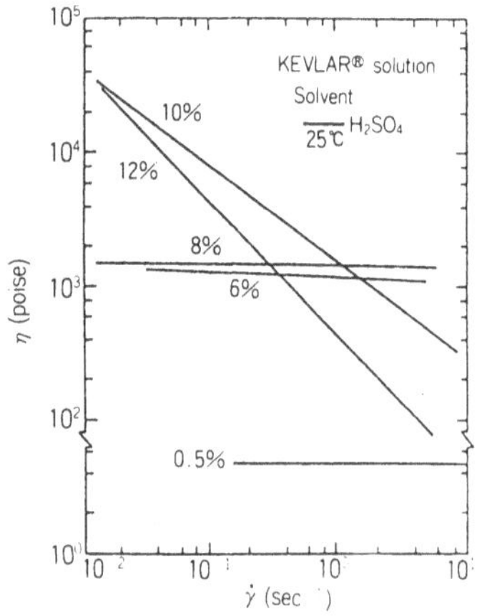

6.7 Shear viscosity vs shear rate for redissolved KEVLAR/H_2SO_4 solution at 25°C.[38]

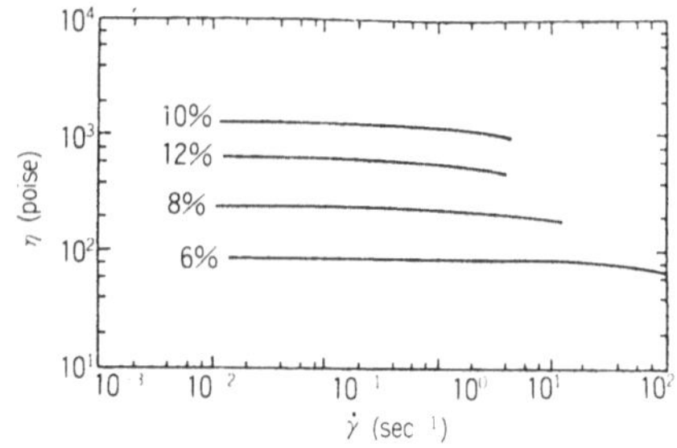

6.8 Shear viscosity vs shear rate for redissolved KEVLAR/H_2SO_4 solution at 60°C.[38]

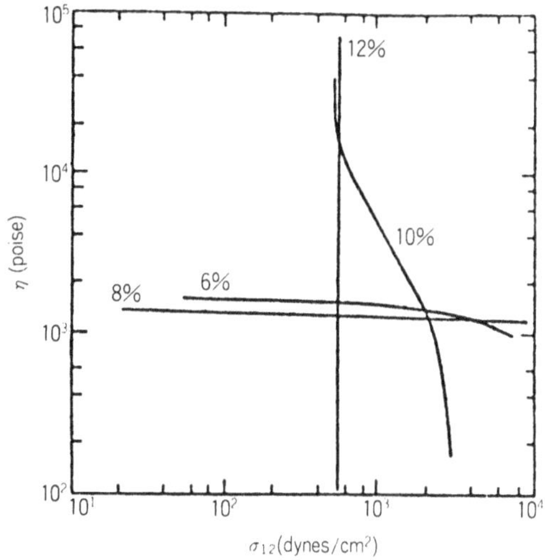

6.9 Shear viscosity vs shear stress of redissolved KEVLAR/H_2SO_4 solution at 25°C.[38]

correlations of shear viscosity, η, against shear rate, $\dot{\gamma}$, for PPD-T/H_2SO_4 solution of various concentrations at 25 and 60°C, respectively. Figure 6.9 is a plot of viscosity against shear stress for PPD-T solutions at 25°C.[38] The changes in slope of these curves between 8 and 10% solutions indicate the effect of isotropic–anisotropic phase transition. The viscosity–shear stress curves for 10 and 12% solutions tend to infinity, indicating the presence of a yield stress.[34]

6.2.3 Fiber preparation

6.2.3.1 Dry-jet wet spinning process

In 1970, Blades[39] discovered that high strength, high modulus fiber could be prepared from the dry-jet wet spinning of anisotropic solutions of aramid polymers. His process is shown schematically in Fig. 6.10. The

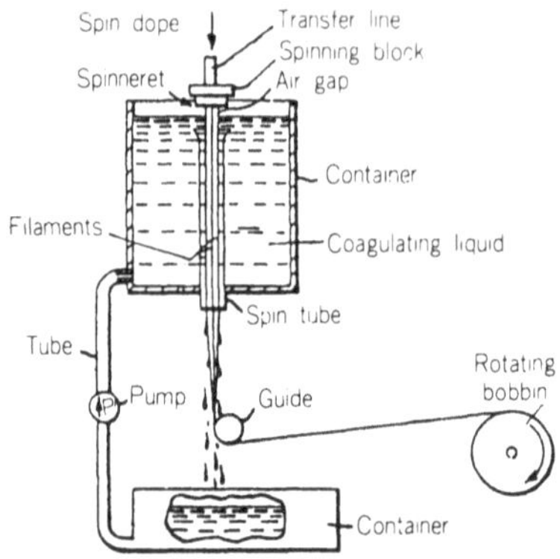

6.10 Dry-jet wet spinning process.[39]

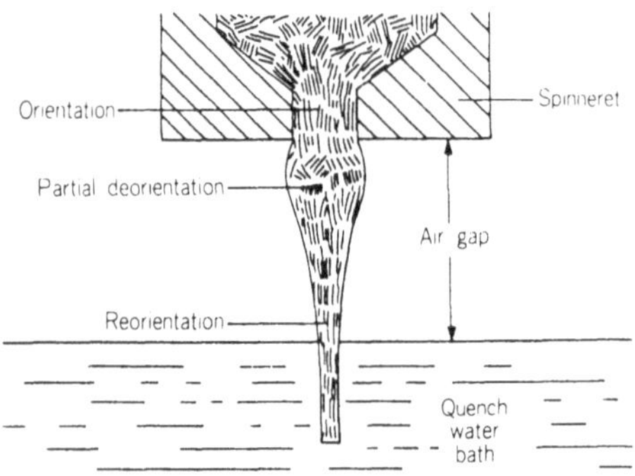

6.11 Molecular orientation during dry-jet wet spinning.[18,40]

key feature of this process is that an anisotropic solution is extruded from spinneret holes through an air gap into a coagulation bath. The coagulated filaments are washed, neutralized and dried. This process produces a fiber with tenacity and initial modulus 2–4 times those of a fiber prepared by the conventional wet spinning process.

The mechanism of polymer molecular orientation in a dry-jet wet spinning process is shown in a conceptual diagram in Fig. 6.11.[40] When an anisotropic solution is extruded through a spinneret capillary, the capillary shear causes the liquid crystalline domains to orient along the direction of flow. At the capillary exit, some deorientation of liquid crystalline domains occurs because of solution viscoelasticity. However, the deorientation is quickly overcome by filament attenuation in the air gap under spinning tension. The attenuated filaments will retain the highly oriented molecular structure when they are precipitated. This gives rise to a highly crystalline, highly oriented fiber structure.

The operating conditions for dry-jet wet spinning are proprietary for fiber producers and are therefore not revealed in detail. A review of literature shows that the general conditions are as follows:

Polymer molecular weight	5000–35 000
Polymer inherent viscosity	3–20 dL/g
Spinning speed	> 55 y/min (> 50 m/min)
Number of filaments	10–1500
Spinneret hole diameter	0.002–0.004 in (0.051–0.102 mm)
Filament size	1–6 denier/filament
Spin stretch factor	< 10

The spin stretch factor is defined as the ratio of windup speed to the filament speed at the spinneret hole.

The dry-jet wet spinning process is quite different from the conventional wet spinning process where the spinning nozzle is immersed in the coagulation liquid. The wet spinning process has several inherent problems. First, the spinning solution must be kept from freezing inside the spinneret by the use of high coagulation temperatures. Secondly, the spinning solution is exposed to the coagulant as soon as it exits the spinneret nozzle. This prevents the solution from complete attenuation. The dry-jet wet spinning method allows the use of low temperature coagulation without freezing the spin solution. The air gap permits the extruded solution to be more fully attenuated and develop a higher degree of molecular orientation.

6.2.3.2 Heat treatment

The as-spun fiber from dry-jet wet spinning can be heat treated at high temperature and high tension to increase its crystallinity and degree of crystalline orientation. The heat treatment conditions are generally in the following ranges:

Temperature	250–550°C
Time	< 10 min
Tension	5–50% of breaking strength

In some cases, particularly with rigid-rod aromatic polyimides, fiber drawing to the order of 2–4*x* is conducted near the glass transition temperature to enhance the fiber tensile properties.

6.2.4 Fiber properties

6.2.4.1 Typical physical properties

Table 6.1 summarizes the key physical properties of several high-strength fibers, including several spun from lyotropic solutions. These fibers are selected from aromatic polyamides, polyimides and heterocyclic polymers. While the physical properties of such fibers are not extensively published, the following ranges are suggested on the basis of current polymer and spinning technology.[18]

Density	1.4–1.6 g/cc
Tenacity	20–40 g/d
Modulus	400–2000 g/d
T_g	typically 350°C
T_m	typically non-melting
T_d	typically > 500°C

where T_g is the glass transition temperature, T_m is the melt temperature, and T_d is the degradation temperature.

Clearly, these physical properties are affected by many factors such as polymer composition, polymer molecular weight, degree of anisotropy of spin solution, spinning and heat treatment conditions. Generally speaking, good tensile properties are derived from polymers of high molecular weight and highly anisotropic spin solutions, and spinning and heat treatment conditions which ensure high degrees of crystallinity and crystalline orientation. Spinning conditions should always be optimized to minimize fiber defects. In practice, this means minimum frictional and hydrodynamic drag, and thorough washing and neutralization where needed.

6.2.4.2 Dry-jet vs wet-jet spinning

For the same polymer, the dry-jet wet spinning process generally gives significantly higher fiber tenacity and modulus and lower elongation than the conventional wet-jet wet spinning process. The difference in fiber tensile properties between dry-jet and wet spinning processes is illustrated in Table 6.3 for several polymers.[18] This comparison is superficial in principle, since the spinning speed, fiber denier, and polymer molecular weight were different.

Table 6.3 Dry-jet vs conventional wet spinning[18]

Polymer	Wet–Jet Spinning[4,5]			Dry–Jet Wet Spinning[6]		
	T	*E*	*M*	*T*	*E*	*M*
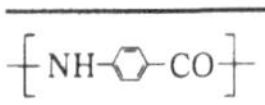	7.2	3.2	350	19	4.0	570
	7.0	9.1	173	26	3.7	750
	4.6	4.8	198	18	6.5	370
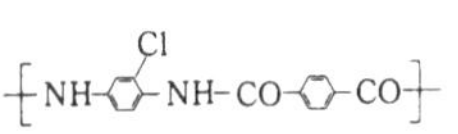	11.4	5.6	379	22	6.9	350
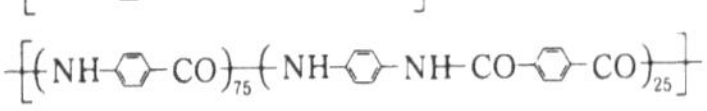	10.2	7.8	264	23	4.8	680

T : tenacity(g/d), E : elongation (%), M : modulus (g/d).

The dry-jet wet spinning process is now widely used for spinning anisotropic, as well as isotropic, solutions. Technora aramid fiber, which is manufactured by Teijin Ltd in Japan, is spun from an isotropic solution via dry-jet wet spinning. As shown in Table 6.1 it offers excellent tensile and physical properties.

6.2.4.3 Effect of polymers

Fibers of lyotropic aromatic polymers exhibit significantly different combinations of tensile properties. The differences in fiber tensile properties are attributable not only to spinning process conditions, but more fundamentally, to the chain conformation of various polymers. For example, many rigid-rod aromatic polyimides and heterocyclic polymers give fibers with much higher modulus and lower elongation at break than semi-rigid aromatic polyamides.

Figure 6.12 presents a correlation of fiber modulus against elongation at break of lyotropic polyamides from published data. Except for a few cases, there is a trend of increasing fiber tenacity with increasing modulus. In general, the tensile properties of these lyotropic fibers fall above the prediction of the Black equation[32] discussed in Section 6.1.5. That is, for a given fiber elongation, the fiber modulus is often greater than that predicted by the Black equation. This is significant in that the fiber modulus achievable by the present spinning technology is far below the theoretical modulus of a given polymer. As for aromatic polyimides and heterocyclic polymers, their as-spun or heat treated fibers have very

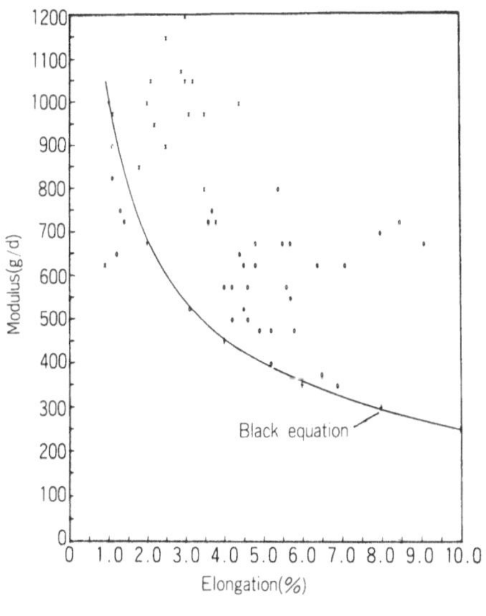

6.12 Modulus vs elongation of lyotropic polymer fibers.

low elongation due to their rigid chain conformation. Their tensile properties deviate greatly from the Black equation.

As with conventional polymers, the molecular weight of lyotropic polymers has a significant effect on fiber tensile properties. Figures 6.13 and 6.14 show correlations of fiber tenacity and modulus against polymer inherent viscosity, respectively.[18] The scattering of experimental data in these correlations is caused by the variety of polymers and the types of spin solution, degree of spin stretch, and post-drawing. For the same polymer inherent viscosity, post-drawing of fibers spun from in situ spin solutions gave significantly higher fiber tenacity and modulus than those

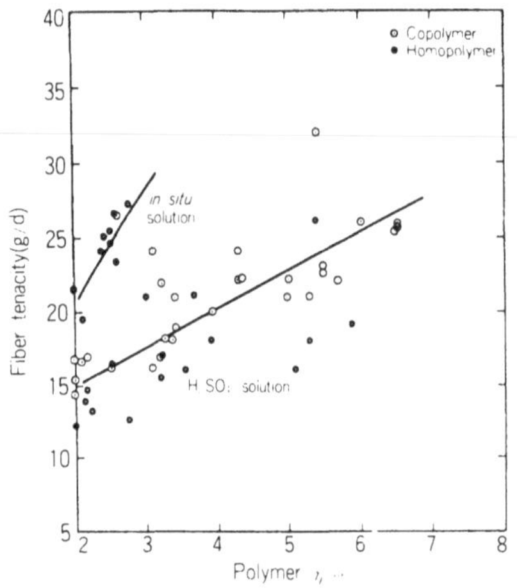

6.13 Fiber tenacity vs polymer inherent viscosity via dry-jet spinning.[18]

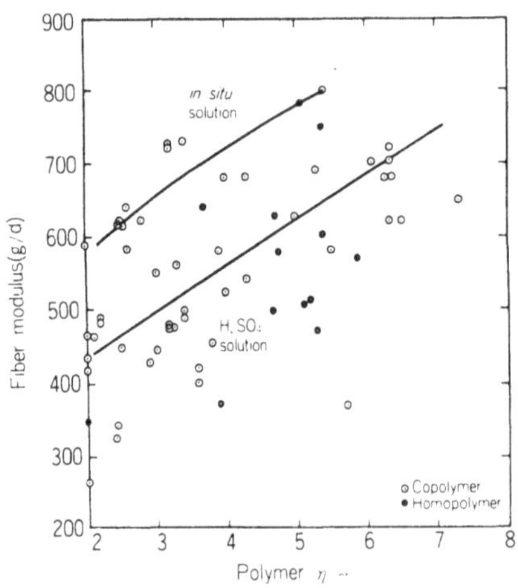

6.14 Fiber modulus vs polymer inherent viscosity via dry-jet spinning.[18]

spun from spin solutions of sulfuric acid. Interestingly, the effect of polymer type, i.e., homopolymer vs copolymer, on fiber properties was fairly small.

6.2.4.4 Fiber strength–modulus behavior

The fibers of lyotropic polymers are characterized by their high strength and high modulus. The key to obtaining such high mechanical properties, on a microscopic scale, is the inherent strength and stiffness of a given polymer by nature of its primary chemical bonding along the chain direction. These mechanical properties are also affected by a weaker secondary interaction between chains because of imperfect crystal orientation. An idealized fiber spinning process, therefore, is to achieve a fully extended chain configuration with maximum uniaxial orientation and lateral packing of a high molecular weight polymer. Although this generally ensures that high fiber strength is accompanied by high fiber modulus, one should expect variations in the fiber strength-modulus relationship due to variations in spinning conditions.

Allen *et al.*[41] studied the deformation and failure behavior of variants of poly (*p*-phenylene terephthalamide) (PPD-T) fibers. They found that fibers of widely different tensile strength and modulus exhibited a similar non-linear elastic deformation. The non-linear behavior was characterized by an asymptotic modulus which described the local crystal orientation and fiber structure at failure. As shown in Fig. 6.15, the asymptotic modulus correlated with the tensile strength of fibers in accordance with shear failure criteria. It was concluded that the tensile failure of these fibers was dominated by the inherent strength anisotropy of the PPD-T polymer and the imperfect orientation in the fibers.

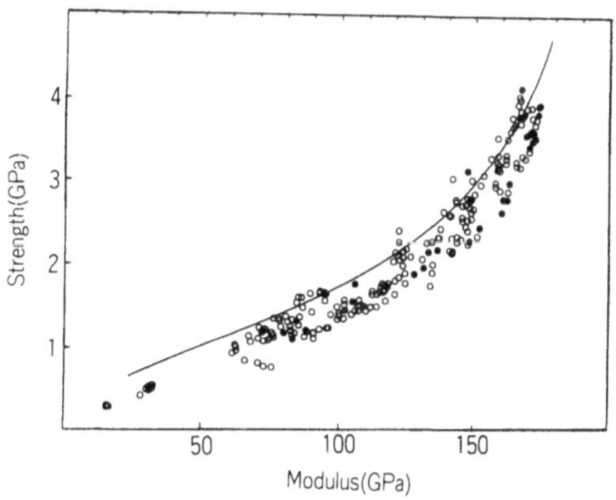

6.15 Tensile strength vs asymptotic modulus of PPD-T fibers.[41]

The strength–modulus correlation of Fig. 6.15 can be generalized for other linear-chain lyotropic polymers. Clearly, one must optimize the spinning process to accommodate a given lyotropic polymer in order to maximize the fiber strength and modulus.

6.3 Thermotropic polymers

6.3.1 Phase behavior

As with lyotropic polymers, a number of physical changes will occur during the phase transitions of thermotropic polymers. These physical phenomena include well-defined melting and freezing temperatures, changes in melt viscosity, density, optical polarization, and microscopic evidence of crystallite formation.

Figure 6.16 illustrates changes in the softening and melting temperatures of aromatic copolyesters of the following general formula:[18,42]

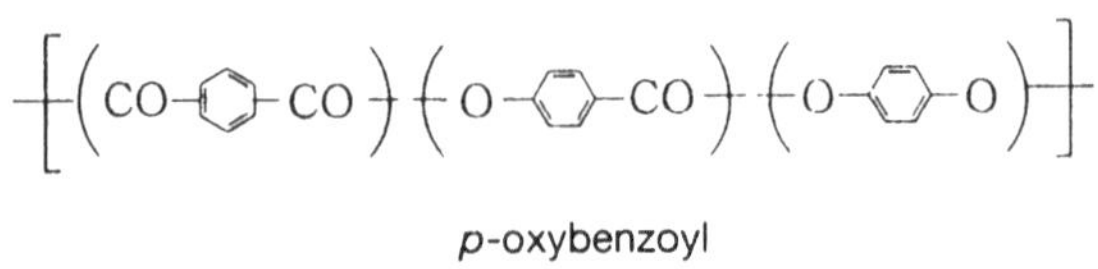

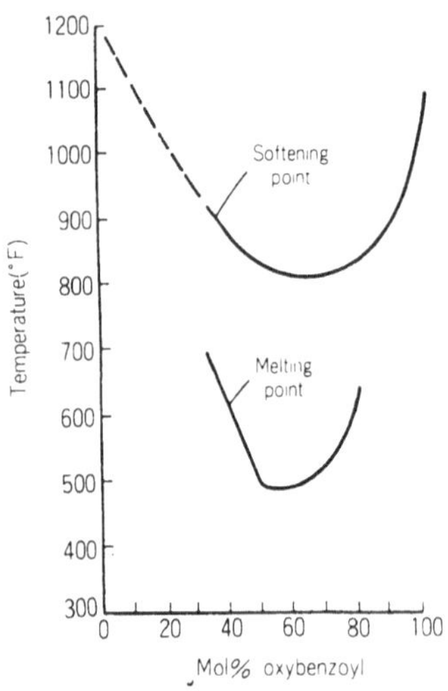

6.16 Softening and melting temperatures of aromatic copolyesters containing *p*-oxybenzoyl unit.[18,42]

The upper curve is for the softening temperature of copolymers containing only para-oriented aromatic ring units. These polymers either were not meltable, or would decompose before reaching their melt temperature. The lower curve is for copolymers containing one meta-oriented aromatic ring unit. These polymers gave reasonable melt temperatures.

The copolyesters exhibited a reduction in the softening and melting temperatures as *p*-oxybenzoyl was increasingly added as a co-ingredient. Both temperatures reached minimum at about 60 mole % *p*-oxybenzoyl units in the copolymer and then increased again with further increase in *p*-oxybenzoyl concentration. Thus, the addition of mesogenic *p*-oxybenzoyl units enhanced the ordering of polymer molecules and the melt anisotropy of this aromatic copolyester. The liquid crystalline phase existed at temperatures slightly above the melt temperature. Above 60% *p*-oxybenzoyl, the melt temperature increased because of increased polymer chain rigidity.

Another example of thermotropic phase behavior is shown in Fig. 6.17 by the correlation of melt viscosity against composition of PET/HBA copolyesters of the following formula:[43]

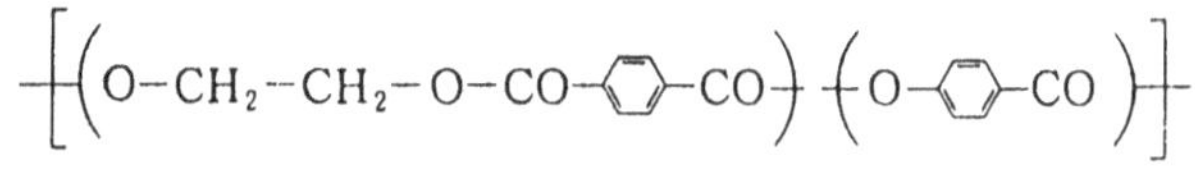

PET/HBA copolyester

This copolyester was prepared from the acidolysis of poly (ethylene terephthalate) (PET) by *p*-acetoxybenzoic acid (HBA) followed by re-polymerisation. The onset of melt anisotropy occurred at about 30 mole

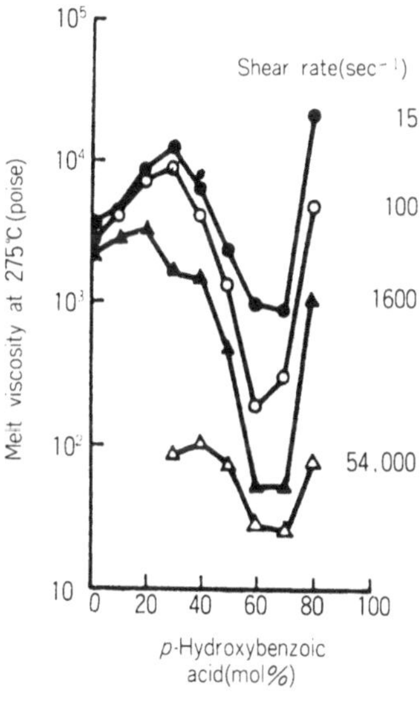

6.17 Melt viscosity of PET/HBA copolyesters.[18,43]

% *p*-oxybenzoyl in the copolymer. The melt viscosity reached a minimum at 60–70 mole % *p*-oxybenzoyl.

6.3.2. Rheological behavior

The melt rheology of thermotropic polymers has not been widely published. Among those polymers which have been reported, the PET/HBA copolyesters have been investigated most extensively. Results of these investigations shed some light on the general behavior of thermotropic polymers.

Wissbrun[44] investigated the rheological behavior of 40/60 PET/HBA copolyester. He observed a yield stress, shear thickening of melt viscosity, sensitivity to temperature, effect of thermal history and absence of die swell. Jerman and Baird[45] made rheological measurements of PET/HBA copolyesters containing 60 and 80 mole % HBA. They also observed the dependence of melt viscosity on temperature and shear rate. Figures 6.18–6.20 show correlations of shear viscosity against shear rate. It is significant to note that the copolyesters gave shear viscosity two orders of magnitude lower than PET. Furthermore, PET behaved differently from the copolyesters in that its shear viscosity was nearly independent of shear rate.

6.3.3 Fiber preparation

6.3.3.1 Melt spinning process

Fibers of thermotropic polymers are produced by the conventional methods. In principle, polymers from melt polymerization can be spun directly into fibers by the melt spinning method. The polymer is first washed with a solvent such as alcohol, acetone, or water, to remove most impurities. It is then melted and extruded through small orifices of a spinning nozzle to form fibers. Using the conventional melt spinning techniques, fibers of thermotropic polymers such as aromatic polyesters can be spun in a speed range of 100–2000 y/min.[46]

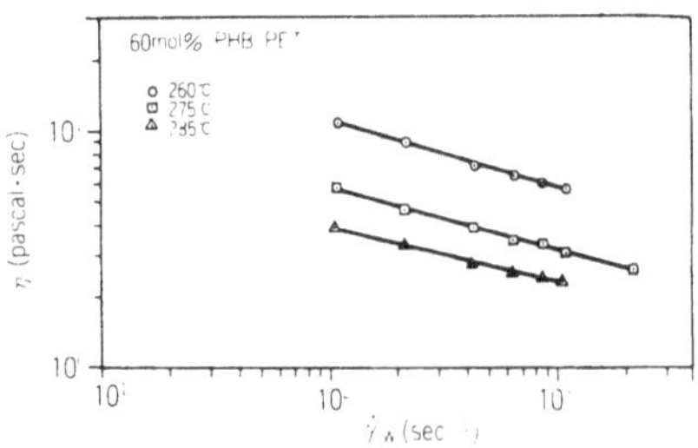

6.18 Viscosity vs shear rate for 40/60 PET/HBA.[45]

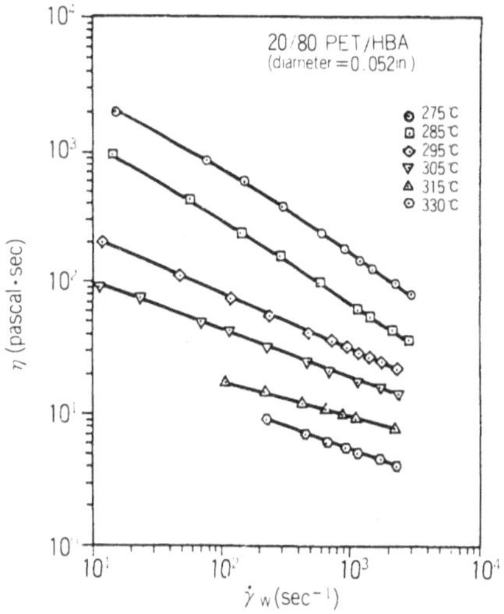

6.19 Viscosity vs shear rate for 20/80 PET/HBA.[45]

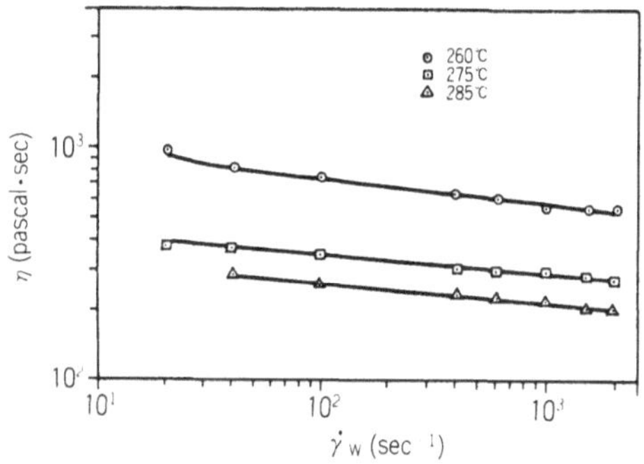

6.20 Viscosity vs shear rate for PET.[45]

During extrusion, the polymer melt must be maintained below the degradation temperature to avoid polymer degradation. Thermotropic polymers which have melt temperatures in the range of 275–375°C and degradation temperatures in the range of 350–450°C, can be processed conveniently without degradation.[47]

Wissbrun and Ide[48] reported that melt processable anisotropic polymers can be processed at reduced temperatures by controlling the thermal history to improve their mechanical properties. They defined a process where an anisotropic melt polymer is subjected to an elevated temperature in the range of 60–85°C above the DSC transition temperature for 0.5 second, and then the melt is cooled by at least 15°C to a processing temperature before converting it into the final product. This process was based on the phenomenon that a substantial decrease in melt viscosity occurs when a liquid crystalline polymer is subjected to an elevated temperature and, upon cooling, retains substantially the reduced melt viscosity for 10–30 minutes. The reduction in melt viscosity, which may be by a factor of 10–100, allows the liquid crystalline polymer to be processed at a lower temperature with improved processability.

The melt flow of thermotropic polymers through small orifices during fiber spinning warrants some consideration. Ide[49] devised a process which subjects an anisotropic polymer melt to elongational flow in order to enhance molecular orientation and improve product properties. A grid plate with tapered holes was inserted in the melt passage leading to the extrusion die or capillary hole. The converging holes may be curves or cone-shaped with a half angle of 20–45° and an exit diameter of approximately 75–150µm. The convergence induces an elongational flow and a stretching motion in the polymer melt along the direction of flow. The preferred elongational rate is in the range of 100–5000 sec^{-1}. Under these conditions, it is theorized that the elongational flow helps reduce the polymer domain structure and allow the polymer molecules to orient more readily during extrusion.

Thus, the general spinning conditions for most thermotropic polymers are as follows.

Polymer inherent viscosity	1.5–3 dL/g
Spinning speed	100–2000 y/min (90–1800 m/min)
Spinning temperature	280–400°C
Number of filaments	50–500
Spinneret hole diameter	0.005–0010 in (0.127–0.254 mm)
Filament size	3–10 denier/filament
Atmosphere	air or inert gas

6.3.3.2 Solution spinning process

Although melt spinning provides a direct route to convert thermotropic polymers to fibers, solution spinning may be desirable for some polymers to avoid excessive melt temperature or high melt viscosity. In some cases, solution spinning is employed to facilitate the solution dispersion of fiber additives such as colorants and flame retardants. In general, the spin solution is prepared by dissolving the polymer in a solvent at a typical concentration of 15–25%.[47] Additives can be conveniently dispersed in the spin solution prior to spinning. The use of spinneret orifices of 0.0025–0.0059 in (0.065–0.15 mm) was specified for solution spinning.

6.3.3.3 Heat treatment

As-spun thermotropic polymer fibers are generally heat treated at elevated temperatures to improve fiber properties. Pletcher[46] first reported the heat treatment of fibers spun from aromatic polyesters. The as-spun fibers were heated in a nitrogen atmosphere at 150–230°C for 3.5–7 hours at little or no tension. The heat treatment temperature was close to the polymer melt temperature in some cases. The time of heat treatment was very long as compared to the conventional annealing process. Since this disclosure, most thermotropic polymer fibers are heat treated as a rule. Yang[18] has shown that the heat treatment conditions generally fall in the following ranges:

Temperature	170–320°C
Time	4–30 hours

These conditions were largely selected from experimental experience in the absence of fundamental understanding. However, it appears that such treatment provides some degree of molecular mobility, and perhaps induces solid-state polymerization and crystal perfection in the fiber. As a result, heat treatment generally improves the physical properties of as-spun fibers.

6.3.4. Fiber properties

Table 6.1 presents the fiber properties of two thermotropic aromatic polyester fibers, Ekonol (a registered trademark of Carborundum Co.) and Vectran (a registered trademark of Hoechst–Celanese Co.). These properties, which are of heat treated fibers, compare favorably with those of lyotropic polymer fibers. Some generalities of as-spun and heat treated fiber properties are discussed below.

6.3.4.1 As-spun fibers

The tensile properties of as-spun thermotropic fibers varied greatly because of variations in polymer composition, molecular weight, spinning and heat treatment conditions. A survey of the literature by Yang[18] (Table 6.4) shows the fiber strength and modulus ranges of as-spun thermotropic fibers.

By and large, the fiber strength and modulus are related to the polymer molecular weight. Figure 6.21 is a plot of as-spun fiber tenacity against inherent viscosity of aromatic polyesters.[18] Since the data base covers a large number of polymer compositions, the same inherent

Table 6.4 Fiber strength and modulus ranges of as-spun thermotropic fibers

	As-spun	Heat treated
Tenacity (g/d)	1–12	3–35
Modulus (g/d)	42–650	480–1200
Elongation (%)	1–3	2–5

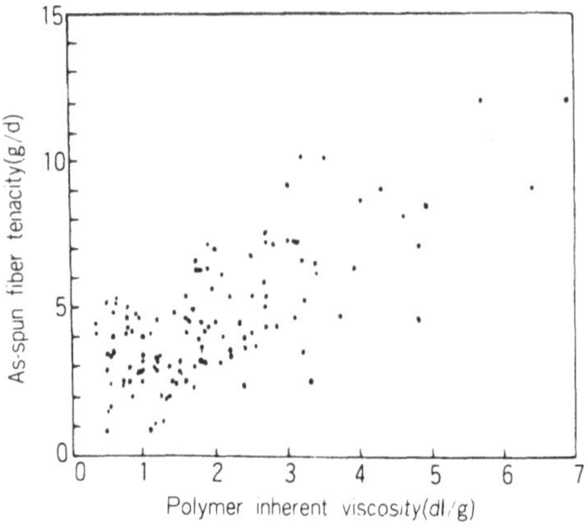

6.21 Effect of polymer inherent viscosity on as-spun fiber tenacity of thermotropic aromatic polyesters.[18]

viscosity does not mean the same molecular weight. Some scatter of data should be expected in this kind of correlation. Nevertheless, the correlation shows a clear trend that fiber tenacity increases with polymer inherent viscosity. Most aromatic polyesters in the literature had inherent viscosities in the range of 1–3. The yarn tenacity from these polymers was in the range of 2–7 g/d.

6.3.4.2 Heat treated fibers

When a thermotropic fiber is subjected to heat treatment, its tenacity generally increases significantly. Figure 6.22 presents a correlation of heat treated fiber tenacity against as-spun fiber tenacity of aromatic polyesters.[18] It can be seen that the tenacity increase due to heat treatment was of the order of 3–4 times that of most fibers. A few fibers did not respond as well and gave only modest tenacity increase. This may be attributed to inadequate heat treatment conditions to enhance post crystallization or to thermal degradation which always depresses yarn strength.

Interestingly, the fiber elongation increased and initial modulus changed very slightly after heat treatment in most cases. These observations are depicted in Fig. 6.23 and 6.24 respectively. Most as-spun fibers had 1–3% elongation, which increased to 2–5% after heat treatment. The initial moduli of most as-spun and heat treated fibers were in the range of 200–500 g/d, and were mostly less than 1000 g/d. It appears that longer heat treatment time and higher temperature tended to give higher yarn elongation. While the increase in fiber tenacity after heat treatment may be attributed to post crystallization, the increase in elongation must result from some form of relaxation. This suggests mechanistically that the heat treatment of aromatic polyester fibers is a rate process. Since the post crystallization and relaxation do not affect

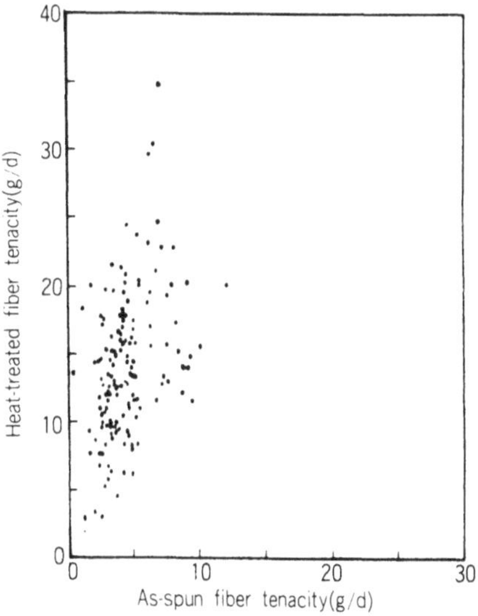

6.22 Effect of heat treatment on fiber tenacity of thermotropic aromatic polyesters.[18]

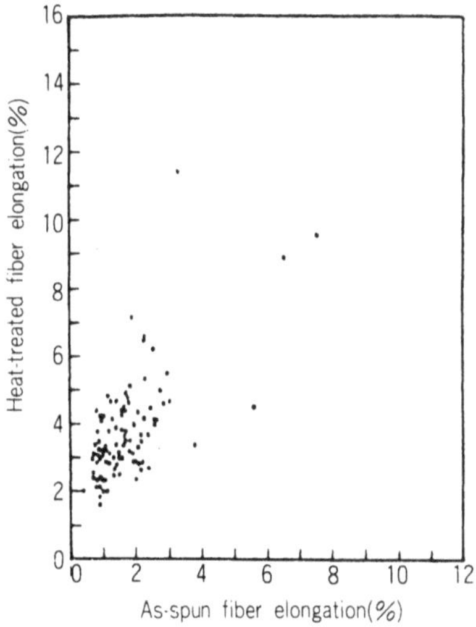

6.23 Effect of heat treatment on fiber elongation of thermotropic aromatic polyesters.[18]

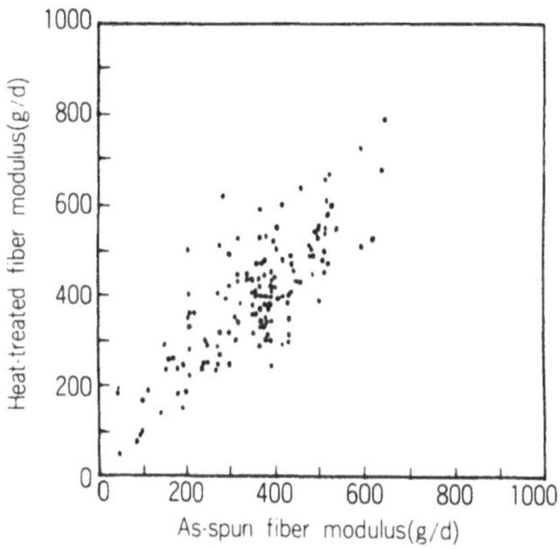

6.24 Effect of heat treatment on fiber modulus of thermotropic aromatic polyesters.[18]

crystal orientation, the yarn modulus is not affected significantly by heat treatment.

Sarlin and Tormala[50] reported recently that heat treatment of thermotropic polymer fibers caused a shift in the DSC melting endotherm. They suggested that the fiber structure was essentially affected by the heat treatment temperature.

Figures 6.25 and 6.26 are correlations of tenacity–modulus and modulus–elongation of heat treated aromatic polyester fibers from published data.[18] As with lyotropic fibers, the tenacity of thermotropic

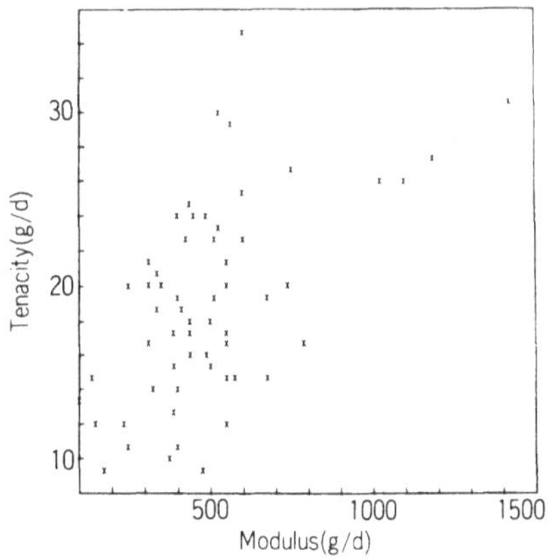

6.25 Tenacity vs modulus of thermotropic aromatic polyester fibers.

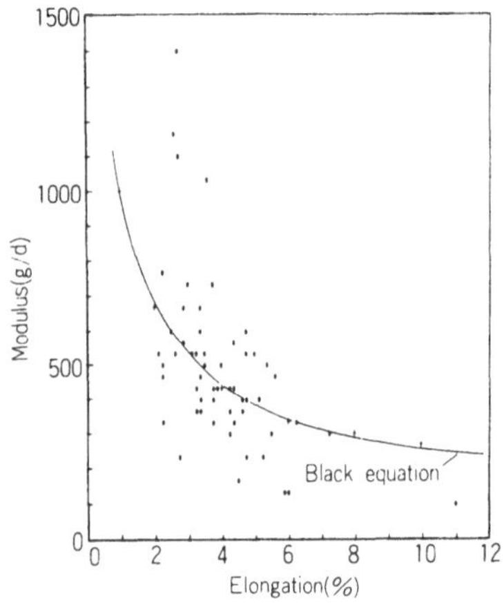

6.26 Modulus vs elongation of thermotropic aromatic polyester fibers.

fibers generally increased with increasing modulus, while elongation decreased. Note that the fiber modulus–elongation relationship followed largely the Black equation in Fig. 6.26.

6.3.4.3 Effect of process conditions

As with other melt-spun fibers, the physical properties of thermotropic fibers are affected by process conditions. Figure 6.27 shows the effect of draw ratio V_f/V_0 on fiber strength of 40/60 and 30/70 PET/HBA copolyesters at an extrusion temperature of 245°C. The fiber strength increased with increasing draw ratio for both copolymers. Figure 6.28 gives a correlation of fiber modulus against draw ratio at various

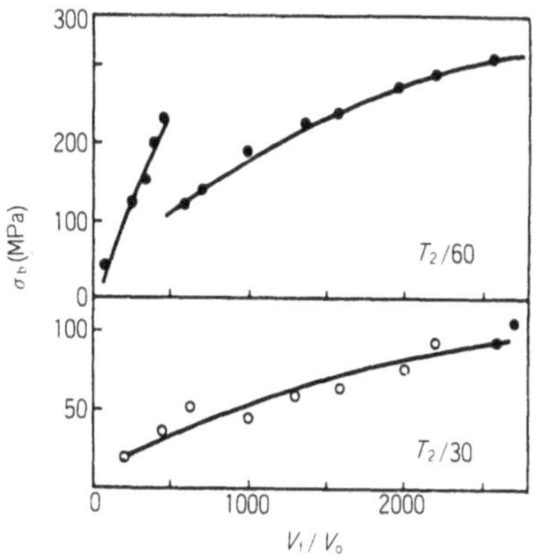

6.27 Fiber strength vs draw ratio V_f/V_0 of 40/60 PET/HBA (top) and 30/70 PET/HBA (bottom) at 245°C extrusion temperature.[51]

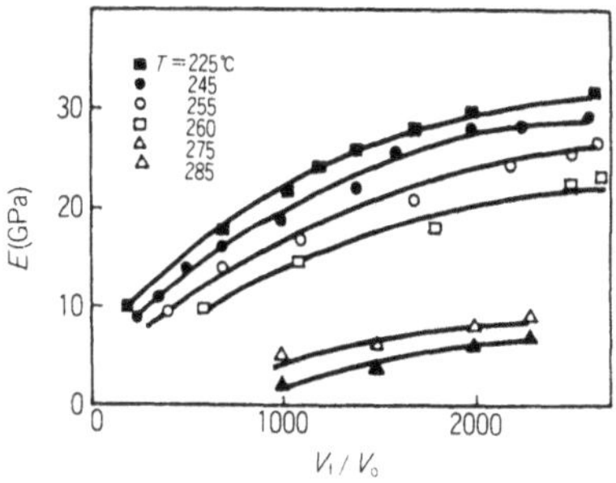

6.28 Fiber modulus vs draw ratio V_f/V_0 of 40/60 PET/HBA at various extrusion temperatures.[51]

extrusion temperatures. The fiber modulus increased with increasing draw ratio, but decreased with increasing extrusion temperature.

6.4 Summary

Anisotropic polymers can generally be converted to high strength, high modulus fibers. Lyotropic polymer fibers are almost exclusively spun from anisotropic solutions by the dry-jet wet spinning process. Thermotropic polymer fibers are mostly spun by the conventional melt spinning process, but can be prepared by the solution spinning process to permit incorporation of additives.

In this chapter, the phase behavior and rheological properties of anisotropic polymer solutions and melts have been discussed. The ranges of basic spinning and heat treatment conditions for both lyotropic and thermotropic polymer fibers have been reviewed. The effects of polymer type, polymer molecular weight, and spinning or drawing conditions on fiber tensile properties were analyzed from published literature.

References

1 Chapoy L L (Ed), *Recent Advances in Liquid Crystalline Polymers*, Applied Science, London, 1985.
2 Gray G W, 'Liquid Crystals and Molecular Structure – Nematics and Cholesterics', in *The Molecular Physics of Liquid Crystals*, Ed G R Luckhurst and G W Gray, Academic Press, New York, 1979.
3 Doi M, *J. Polym. Sci. Polym. Phys. Ed.*, **19**, 224, 1981.
4 Moscicki J K, 'Advances in Chemical Physics', **LXIII**, 631-717, 1985.
5 Brown G H, *Anal. Chem.*, **41**, 26A, 1969.
6 Schaefgen J R, Bair T I, Ballou J W, Kwolek S L, Morgan P W, Panar M, and Zimmerman J, 'Rigid Chain Polymers: Properties of Solution and Fibers', in *Ultra-High Modulus Polymers*, Ed A Ciferri and I M Ward, Applied Science, London, 173, 1977.
7 Charley C L, and Porter R S, *J. Rheology*, **28**, 249, 1984; **29**, 281, 1985.
8 Chung T S, *Polym. Eng. Sci.*, **26**, 901, 1986.
9 Wissbrun K F, *J. Rheology*, **25**, 619, 1981.
10 Baird D G, *Polym. Sci. Technol.: Vol. 28, Polymeric Liquid Crystals*, Proc. Second Symp., ACS meeting, Washington, DC, 1985, Ed A Blumstein, Academic Press, 1985.
11 Kulichikhin V G, Malkin A Ya and Papkov S P, *Polym. Sci. – USSR*, **26**, 499, 1984.
12 White J L, *J. Appl. Polym. Sci., Appl. Polym Symposium*, **41**, 241, 1985.
13 Onogi S and Asada T, Rheology Proceedings Intl. Congress, **1**, 127, 1980.
14 Chu S G, Venkatraman S, Berry G and Einaga Y, *Macromolecules*, **14**, 939, 1981.
15 Calundann G W and Jaffe M, Proceedings of the Robert A. Welsh Conference on Chemical Research, XXVI, Synthetic Polymers, 1982.
16 Metzner A B and Prilutski G M, *J. Rheology*, **30**, 661, 1986.
17 Kwolek S L, 'Liquid Crystalline Polymers', in *Encyclopedia of Polymer Science & Technology*, Vol. 9, 2nd edn, Wiley, 1987.
18 Yang H H, *Aromatic High-Strength Fibers*, Wiley Interscience, New York, 1989.
19 Kwolek S L, *Am. Inst. Chem., The Chemist, Washington, DC*, **57**, 9, 1980.
20 Kwolek S L, Morgan P W and Sorenson W R (Du Pont), US Pat. 3,063,966, 1962.
21 Kwolek S L (Du Pont), US Pat. 3,671,542, 1972; reissued US Pat. 3,819,587, 1974.
22 Morgan P W and Pletcher T C (Du Pont), US Pat. 3,869,419, 1975.
23 Gulrich L W and Morgan P W (Du Pont), US Pat. 3,836,498, 1974.
24 Morgan P W (Du Pont), US Pat. 3,827,998, 1974.
25 Bair T I and Morgan P W (Du Pont), US Pat. 3,673,143, 1972; US Pat. 3,817,941, 1974.
26 Sackman H and Demus D, *Mol. Cryst. Liq. Cryst.*, **21**, 239, 1973.
27 de Vries A, *Mol. Cryst. Liq. Cryst.* **24**, 337, 1973.

28 Flory P J, *Proc. Roy. Soc. (London)*, Series A, **234**, 73, 1956.
29 Benoit H and Doty P, *J. Phys. Chem.*, **57**, 958, 1953.
30 Ying B and Chu B, *Makrom. Chem., Rapid Comm.*, **5**, 785, 1984.
31 Tanford C, *Physical Chemistry of Macromolecules*, Wiley, New York, 1963.
32 Black W B, *J. Macromol. Sci., Chem.*, **A7**, 149, 1973.
33 Hancock T, Spruiell J E and White J L, *J. Appl. Polym. Sci.*, **21**, 1227, 1977.
34 Papkov S P, Kulichikin V G, Kalymykovo V D, and Malkin A Ya, *J. Polym. Sci.*, **12**, 1953, 1974.
35 Aoki H, Coffin D R, Hancock T A, Harwood D, Lenk R S, Fellers J F and White J L, *J. Polym. Sci., Polym. Symp.*, **65**, 29, 1979.
36 Baird D G, *J. Appl. Polym. Sci.*, **22**, 2701, 1978.
37 Baird D G and Ballman R L, *J. Rheology*, **23**, 505, 1979.
38 Aoki H, White J L and Fellers J F, *J. Appl. Polym. Sci.*, **23**, 2293, 1979.
39 Blades (Du Pont), US Pat. 3,767,756, 1973; US Pat. 3,869,429, 1975.
40 Yang H H, 'Aramid Fibers', in *Fiber Reinforcement for Composite Materials*, Ed A R Bunsell, Elsevier Science Publishers, Amsterdam, 1988.
41 Allen S R, Roche E J, Bennett B and Molaison R, *Polymer*, **33**, 1849, 1992.
42 Cottis S G, Economy J and Nowak B E, US Pat. 3,637,595, 1972.
43 Jackson W J and Kuhfuss H F, *J. Polym. Sci., Polym. Chem. Ed.*, **14**, 2043, 1976.
44 Wissbrun K F, *Br. Polym. J.*, **12**, 163, 1980.
45 Jerman R E and Baird D G, *J. Rheology*, **25**, 275, 1981.
46 Pletcher T C (Du Pont), US Pat. 3,991,013, 1976.
47 Williams A G, US Pat. 4,354,994, 1982.
48 Wissbrun K F and Ide U (Celanese), US Pat. 4,325,903, 1982.
49 Ide Y (Celanese), US Pat. 4,332,759, 1982.
50 Sarlin J and Tormala P, *J. Polym. Sci., Part B: Polym. Phys.*, **29**, 395, 1991.
51 Acierno D, LaMantia F P, Polizzotti G, Ciferri A and Valenti B, *Macromolecules*, **15**, 1455, 1982.

7

Spinning of Thermotropic Liquid-Crystal Polymers

Junyou Nakagawa
Kuraray Ltd, Kurashiki, Japan

7.1 Introduction

Liquid-crystal polymers include the lyotropic liquid-crystal type and the thermotropic liquid-crystal type. Pioneering studies on the latter were carried out by Economy[1] of Carborundum Co., Jackson[2] of Eastman Kodak and Calundann[3] of Hoechst–Celanese, together with their collaborators, and some of the polymers are now commercially available as resins. Table 7.1 shows various thermotropic liquid-crystal polymers on the market, and their suppliers.[4]

In the fiber field, however, although numerous studies have been proceeding since 1970, 'VECTRAN' is the only fiber that is currently

Table 7.1 Commercial thermotropic liquid-crystal polymer resins

Type	Chemical structure (determined)	Supplier (trade name)
Flexible group-introduction type	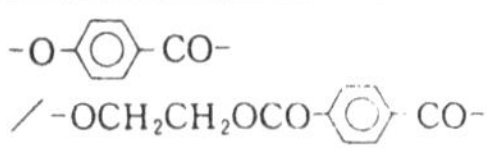	Unitika (RODRAN) Mitsubishi Chemical Ind. (NOVACCURATE)
Aromatic ring substitution type	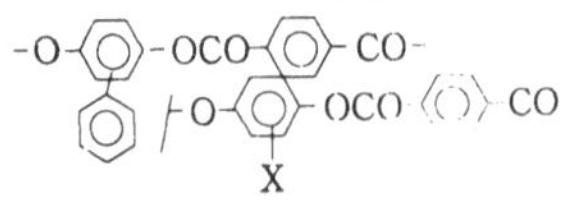	Du Pont (HX) Granmont (GRANLAR)
Rod-like molecule copolymerisation type	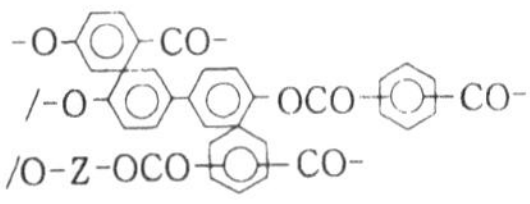	Amoco (XYDAR) Nippon Oil (XYDAR) Sumitomo Chemical (EKONOL) Toso (HAG, HBG)
	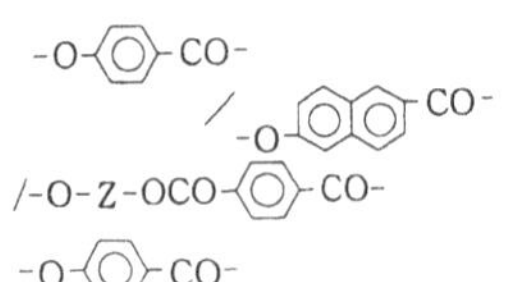	Hoechst–Celanese (VECTRA) Polyplastics (VECTRA) Ueno Fine Chemicals Ind. (UENO LCP)
	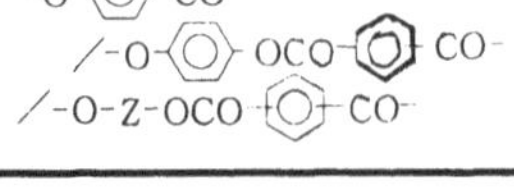	BASF (ULTRAX) Bayer (POLYSTAL) ICI (VICTREX SRP)

commercially available. Kuraray, using VECTRA polymer and fiber-forming techniques formulated by Hoechst–Celanese, succeeded in commercialising the fiber.

Melt spinning is the most rational fiber-formation process for the polymer. Continuing research and development work may yet achieve super-fibers having higher performance and characteristics at lower cost.

7.2 Thermotropic liquid-crystal polymers

To obtain a high-strength, high-modulus fiber, it is necessary to select a polymer having a molecular structure in which molecular chains are packed in a small cross-sectional area, and having strong bonds and low elongation. Desirable examples for this purpose are aromatic polymers comprising a highly symmetrical and rigid rod-like structure. With wholly aromatic polyesters, the basic structures include a self-condensation polyester of (a) *p*-hydroxybenzoic acid (HBA) and (b) a polyester comprising units from terephthalic acid (TA) and units from hydroquinone (HQ).

(a)

(b)

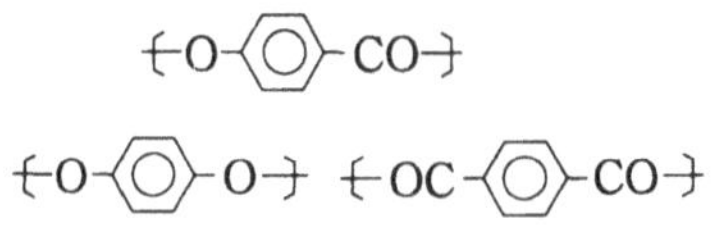

The above polymers, however, have melting points ((a): 610°C, (b): 600°C) higher than their decomposition temperatures (400–450°C) due to their molecules being too rigid, and therefore are unprocessable in the thermotropic liquid crystal form. To overcome the problem, attempts have been made by various methods to decrease the melting point to below the decomposition temperature at the expense, to some extent, of the rigidity and crystallinity of the polymers. The methods include:

1 The introduction of units from a flexible alkyl group into the main chain.
2 Introduction of a substituent into each of the aromatic rings in the main chain.
3 Copolymerization of two rigid molecules.
4 Introduction of a rod-like non-linear structure.

These methods have been able to provide a series of thermotropic liquid-crystal polymers that are injection-moldable or melt-spinnable, which will be described later.

In practice, thermotropic liquid-crystal polymers are obtained by polymerization of an aromatic diol acetate and an aromatic dicarboxylic acid or self-condensation of an aromatic acetoxycarboxylic acid.

7.3 Spinning of thermotropic liquid-crystal polymers

Liquid crystals have a flowable, optically anisotropic structure above a certain temperature and lack, in this temperature region, sufficient energy for their individual molecules to rotate freely as in a liquid. Consequently, many inter-molecular interactions occur in the liquid and there form domains comprised of parallel-arranged molecules. Nematic liquid crystals generally form, with an orientation direction parallel to the long axes of the molecules present. In terms of processability, thermotropic liquid-crystal polymers generally allow the molecules to be oriented when subjected to shear stress and also exhibit long orientation relaxation times. These are the main features that distinguish them from conventional flexible polymers. In a transition flow study of thermotropic liquid-crystal polymers, relaxation of stress can last seconds, whereas the relaxation of orientation can be a matter of several minutes[5]. When a thermotropic liquid-crystal polymer is extruded through an orifice under a high shear stress, its molecules orient to a high degree and the oriented structure is almost completely maintained during solidification by cooling due to the long relaxation time. From a technical viewpoint, the following three processes are available for spinning thermotropic liquid-crystal polymers:

1 Spinning of polymers having a comparatively low molecular weight (low η_{inh}).
2 Spinning of high molecular weight polymers at a temperature higher than their melting point.
3 Spinning of high molecular weight polymers at a temperature lower than their melting point (super-cooled spinning).

While operation of process 1 is rather easy, the as-spun fiber obtained has a low strength and hence requires a protracted heat treatment for achieving high performance.

With process 2, the molecular weight is increased to the maximum level that still assures spinnability and the spinning is conducted at a high temperature and high shear rate to decrease melt viscosity. The as-spun fiber obtained has a strength of 5–20 g/d which can be further improved by a relatively short heat treatment. The thermal stability of the polymer used and spinning skill are the key points in this process.

Process 3 was studied by Zimmerman[6] and comprises extruding a high molecular weight liquid-crystal polymer having a ΔH_c of not more than 10 J/g at a temperature not more than the melting point and not less than the freezing point and taking up the as-spun fiber at a relatively low speed. This process is of interest, because the polymer thermally decomposes only slightly and the as-spun fiber obtained has a high strength and requires only short heat-treatment time. The spinning of thermotropic liquid-crystal polymers is described later in more detail.

7.4 Heat treatment

The as-spun fiber obtained by melt spinning a thermotropic liquid-crystal polymer already has a high strength and modulus. To improve the performance still further, the as-spun fiber is subjected to heat treatment, which improves the strength and modulus, and also the thermal resistance. The as-spun liquid-crystal polymer fiber is generally heat treated in a reduced-pressure atmosphere of an inert gas, or air, with the by-products that form being continuously removed. The heat treatment is conducted at a temperature not more than the melting point, to avoid the filaments sticking. Since the melting point itself increases during heat treatment, the treatment temperature can eventually be higher than the original melting point of the polymer. As will be described later, it has been confirmed that heat treatment causes solid-phase polymerization, thereby increasing the molecular weight. The reaction mechanism is, however, complex and there may possibly occur crosslinking reactions and also elongation of molecular chains. The fiber formation of thermotropic liquid-crystal polymers is schematically illustrated, in comparison with flexible polymers, in Fig. 7.1.

7.5 Fiber formation and fiber characteristics

7.5.1 Fibers of the flexible group-introduction type

A copolyester obtained by polymerising 60 mol% of HBA and 40 mol% of polyethylene terephthalate (hereafter referred to as PET/HBA:40/60) is a raw material of commercial interest from the viewpoint of production cost. This polymer however has not yielded particularly good results as a material for fiber, as reported by Acierno *et al.*[7] and Jackson.[8] On the

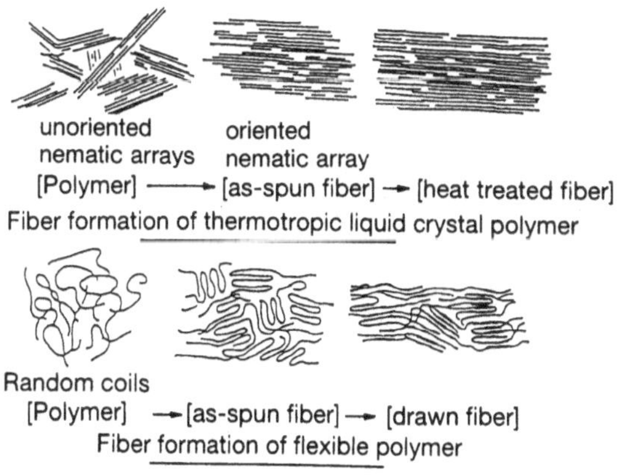

7.1 Schematic diagram of fiber formation

other hand, with PET/HBA:20/80, Zachariades and Logan report that this copolymer was spun at a high temperature of 330°C to give a fiber having a strength of 225 MPa and a modulus of 40 GPa.[9] While PET/HBA:40/60 is a random copolymer,[10] 20/80 is considered to contain blocks of HBA.[11] According to a patent, irregular particles formed upon block-wise polymerization of HBA were removed and the copolymer thus homogenized was spun at 310°C to give a fiber having a strength of 10.1 g/d, which was then heat treated at 170–265°C for 4 hours, to show an improved strength of 25.2 g/d.[12] Jackson used, instead of PET, a polymer from 3,4′-carboxybenzenepropionic acid (CPA) and HQ. A copolyester of CPA-HQ/HBA:20/80 was spun at a temperature of 335°C and at a take-up speed of 1070 m/min, to obtain an as-spun fiber having a strength of 6.9 g/d and a modulus of 470 g/d, which they further heat-treated at 325°C for 0.5 hour to obtain a fiber having a strength of 20.3 g/d and a modulus of 960 g/d.[13]

7.5.2 Fibers of rod-like molecule-copolymerization type

Fiber formation from a copolymer of HBA and 2,6 hydroxynaphthoic acid (HNA), which is commercially employed for VECTRAN, is explained here. X-ray analysis by Gutierrez *et al.* has revealed that the polymer of HBA/HNA is a random copolymer.[14] Calundann and Jaffe[3] spun a polymer of HBA/HNA:75/25 having an η_{inh} (inherent viscosity as determined in pentafluorophenol) of 5.7 dl/g and a melting point of 302°C, to obtain an as-spun fiber having a strength of 12.1 g/d, elongation of 2.8% and modulus of 541 g/d. The polymer had a comparatively high molecular weight and was spun at a temperature well above the melting point.

Upon melting, hydrolysis and thermal decomposition of the polymer occur, degrading the polymer and generating gases that markedly decrease the spinnability. To minimize the decrease in η_{inh} during spinning, it becomes necessary to lower the moisture content of pellets and to decrease the spinning temperature as far as possible. Figure 7.2 shows the relationship between the rate of shear τ and the melt viscosity.[3] Flexible polymers such as PET generally maintain their viscosity in low-shear regions. On the other hand, with thermotropic liquid-crystal polymers, the viscosity falls nearly linearly as shear rate rises, thus showing marked non-Newtonian behaviour, and is influenced to a large extent by temperature changes.[15]

Fiber formation from thermotropic liquid-crystal polymers includes no drawing process, and so the fineness (denier) of the as-spun fiber becomes that of the finished fiber. To obtain a fine-denier fiber, it is necessary either to use a small-diameter orifice or to increase the spinning draft.

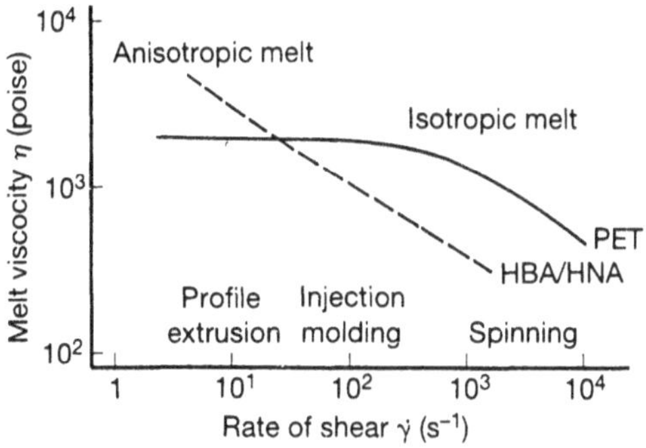

7.2 Relationship between shear rate and melt viscosity

However, as seen from Fig. 7.2, when a thermotropic liquid-crystal polymer passes through an orifice, leaving behind a low τ region, molecular orientation occurs to a high degree, thereby rendering it impossible to employ a large spinning draft. Consequently, to obtain a fine-denier fiber, the orifice diameter must be made small, and as a result τ increases. A large τ in turn causes the melt viscosity to decrease, which is favourable for spinning high η_{inh} polymers.

Figure 7.3 shows examples of the attenuation behavior of extruded streams. PET, which is a flexible polymer, attenuates rapidly down to a diameter of 30 cm from the orifice and then only gradually until taken up. On the other hand, HBA/HNA attenuates rapidly just below the orifice and the attenuation is nearly complete at a distance of 10 cm from the orifice. Calundann and Jaffe also state that solidification of extruded streams occurs rapidly and that orientation and structure formation are complete within 10 cm from the orifice.[3] Under the usual spinning conditions, no 'die-swell' phenomenon is observed. Krigbaum *et al.*

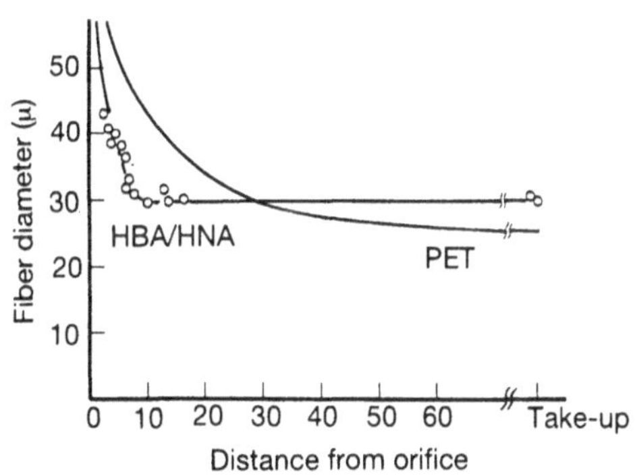

7.3 Attenuation behaviour of extruded streams.

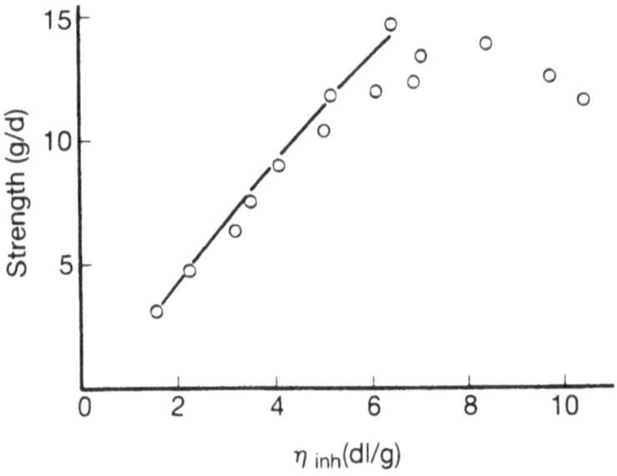

7.4 Relationship between η_{inh} and strength of as-spun fiber

carried out spinning of HBA/HNA:75/25 and reported that the die swell is a function of shear stress and is not observed above a shear stress of about 8 x 10^4 dyn/cm.[16]

Figure 7.4 shows the results obtained by Calundann and Jaffe for the relationship between η_{inh} and strength of as-spun fiber.[3] The strength reaches a maximum of nearly 15 g/d at an η_{inh} of about 7 dl/g, but then decreases because stable spinning becomes impossible due to high viscosity. Good spinnability is obtained with a slightly lower η_{inh}, which gives an as-spun fiber having a strength, elongation and modulus of about 12 g/d, 2% and 600 g/d, respectively.

Heat treatment is conducted at a temperature below the melting point to further improve the properties. Figure 7.5 shows DSC curves of heat-treated fibers. The endothermic melting peak shifts toward higher temperature and has a smaller area with increasing treating temperature. A fiber heat-treated until the endothermic peak temperature exceeds about 300°C is no longer meltable under atmospheric pressure and, when heated at higher temperatures, gradually decomposes while keeping its fiber shape. The heat treatment time required to give sufficient improvement in properties and the strength achieved may vary depending on the manufacturing conditions of the polymer and the way in which the heat treatment is conducted. Yoon reported that a polymer of HBA/HNA:73/27 was spun and the as-spun fiber was heat treated at 270°C for 30 minutes, to give a fiber having a strength of about 3.5 GPa, which eventually reached about 3.71 GPa (29.8 g/d).[17]

The strength of as-spun fibers comprising HBA/HNA markedly increases upon heat treatment, but the modulus increases only slightly. An as-spun fiber spun from a copolymer obtained by copolymerization of HBA/HNA with 4,4-biphenol (BP) and terephthalic acid (TA), with reduced amount of HNA, i.e. HBA/HNA/BP/TA: 60/5/17.5/17.5

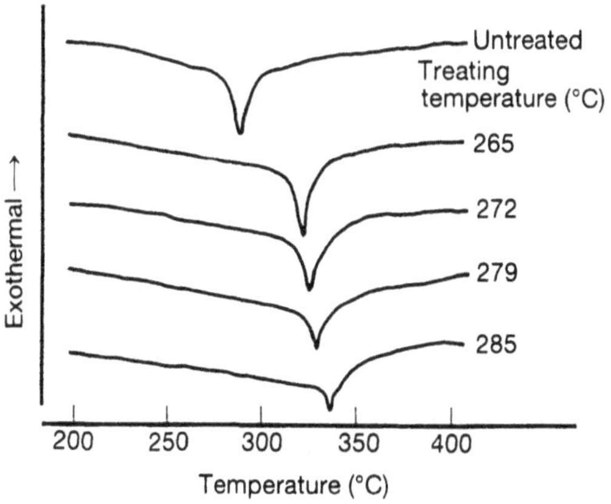

7.5 DSC curves of heat-treated fibers

improves, on heat treatment at 300 to 310°C, not only in strength but also in modulus from 634 up to 1110 g/d,[18] as shown in Table 7.2.

Fibre formation from EKONOL (i.e. poly HBA)-related liquid-crystal polymers was studied by Economy *et al.*[24] and these products have since been promoted by Sumitomo Chemical Co. up to the stage of development under the name of 'EKONOL FIBRE'.[25] Sumitomo spun a copolymer having a non-linear (rigid-angular) component, isophthalic acid

Table 7.2 Composition of liquid-crystal polymers and fiber properties

Polymer				As-spun fiber		Heat-treated fiber		Reference
Type*	Composition	Ratio (mol%)	η_{inh}	DT	IM	DT	IM	
A	PET/HBA	40/60	–	3.3	196	–	–	8
	PET/HBA	40/60	–	10.1	–	25.2	–	12
	CPA-HQ/HBA	20/80	–	6.9	470	20.3	960	13
B	HBA/HNA	75/25	5.7	12.1	541	20.0	550	19
	HBA/HNA	73/27				29.8		17
	HBA/HNA/TA/BP	60/5/17.5/17.5	6.5	7.9	634	27.8	1110	18
	HBA/HNA/TA/BP	60/5/15/20	–	5.5	410	30.8	1420	20
C	PhHQ-TA	100	3.0	3.4	397	32.0	910	13
	PhHQ-TA/HBA	83.8/16.2	2.4	6.1	400	34.8	485	21
	PhHQ-TA/BP-TA/HBA	80/10/10	9.3	20.1	–	–	–	22
	ClHQ-TA/ClHQ-NA	70/30	2.6	6.1	547	30.4	527	23

NA: 2,6-naphthalenedicarboxylic acid, PhHQ: phenylhydroquinone, DT: dry tensile strength (g/d), IM: initial modulus (g/d).
*A Flexible group-introduction type.
B: Rodlike molecule-copolymerization type.
C: Aromatic ring-substitution type.

(IP) of composition HBA/IP/TA/BP:60/5/15/20, at a temperature of 350°C and at a take-up speed of 160 m/min to obtain an as-spun fiber having a strength, elongation and modulus of 5.5 g/d, 1.4% and 410 g/d, respectively. When heat treated the as-spun fiber achieved a strength and modulus of 30.8 g/d and 1420 g/d, respectively.[20]

7.5.3 *Fibers of aromatic ring-substitution type*

The homopolymer of phenylhydroquinone (PhHQ) and TA has a melting point of 341°C and is melt-spinnable. Jackson *et al.* heat treated an as-spun fiber of this polymer having a fineness, strength and modulus of 1.2 d, 4.8 g/d and 500 g/d, respectively, at 340°C for 1 hour to obtain a fiber having a strength and modulus of 32 g/d and 910 g/d, respectively.[25] Copolymers have lower melting points and hence even those having high molecular weight (high η_{inh}) are spinnable. It is worth noting that a polymer of PhHQ-TA/BP-TA/HBA:80/10/10 having a high η_{inh} of 9.3 dl/g has a melting point of 288°C and gives, upon spinning at 324°C, an as-spun fiber having a remarkably high strength of 20.1 g/d.[22]

Zimmerman reports that while draw resonance often occurs in the spinning of polymers having low heats of fusion, ΔH_c , of not more than 10 J/g, spinning them at a temperature below their melting point and above their freezing point, i.e. super-cooled spinning, can eliminate draw resonance, thereby rendering it possible to conduct stable spinning.[6] For example, a polymer of PhHQ-TA/HNA:96/4 having an η_{inh} of 6.9 dl/g has a melting point of 348°C and a ΔH_c of 2.1 J/g. Spinning this polymer at 350°C causes serious draw resonance so that the extruded streams are difficult to take up. Spinning at 324°C, however, assures stable take-up, to obtain an as-spun fiber having a strength of 15.6 g/d.

Further heat treatment gives, for example, from a copolymer of PhHQ-TA/HBA:83.8/16.2, a fiber having a strength of 34.8 g/d[26] and from a copolymer using chlorohydroquinone, i.e. CIHQ-TA/CIHQ-NA:70/30, a fiber of 30.4 g/d.[23]

7.6 Fiber structure

The structure of thermotropic liquid-crystal polymer fibers varies depending on the raw material polymer and fiber manufacturing conditions but generally is a highly oriented fibrillar structure. With HBA/HNA fiber, macrofibrils having a size of about 5μm, fibrils having a size of about 0.5μm and microfibrils having a size of about 0.05μm, together with a very thin skin layer having a thickness of about 1μm have been observed.[27] Donald *et al.* observed a banded structure in polarized light, with the striations lying perpendicular to the fiber axis. This structure closely resembles those previously reported in lyotropic polymers.[26]

Table 7.3 shows how the structural factors change upon heat treatment. Thus, the principal dispersion temperature increases, the degree of orientation shows almost no change and the crystallinity obtained from specific weight or by X-ray analysis increases only a little. On the other hand, the number average molecular weight increases by about 3 times (Table 7.4).

Xiao and Takahashi pointed out that while the arrangement of HBA and HNA in the molecular chain does not change, the *a*-axis and *b*-axis X-ray data do change, which suggests that sequences rich in HBA may have aggregated.[28] Yang and Krigbaum identified block formation from HNA caused by melt interchange.[29] Erdemir *et al.* reported the results of an X-ray study, based on changing the composition of a liquid-crystal polymer comprising HBA/HQ-IP, of the structural changes between unoriented polymer, as-spun fiber and heat-treated fiber. According to their study, as-spun fiber from the polymer having a composition of 50/50 has a slightly oriented para-crystalline structure, but there forms, upon heat treatment, a new crystal structure which is different from those of the corresponding homopolymers.[30]

It is interesting that, as described in the previous section, there is a fiber type which, upon heat treatment, shows a marked increase in melting point and strength but shows little change in the modulus and another type which shows a marked increase, not only in melting point and strength, but also in modulus. A representative example of the former is HBA/HNA, which changes little in orientation and crystallinity upon heat treatment. Examples of the latter are HBA/HNA/BP/TA and HBA/IP/BP/TA, in which HNA or IP is only a minor component. With respect to the structural change due to heat treatment, which is very interesting, much remains to be studied.

Table 7.3 Change of structural factors upon heat treatment

	As-spun fiber	Heat treated fiber
α-transition temperature Tα (°C)	88	97
tan δ (max)	0.104	0.105
Sonic velocity (km/s)	8.64	8.65
Crystal orientation (°C)	88	88
Specific weight	1.40	1.41
X-ray crystallinity (%)	45	46

Table 7.4 Change of molecular weight upon heat treatment

	$M_n/10^4$	$M_w/10^4$	$M_z/10^4$
As-spun fiber	1.63	3.84	6.88
Heat-treated fiber	4.89	14.5	27.5

7.7 Conclusion

A number of studies have been carried out on thermotropic liquid-crystal polymer fibers. At present, however, VECTRAN is the only fiber commercially available. From now on, studies will place more emphasis on the development of commercial production techniques for other thermotropic liquid-crystal polymers. The most pressing problem is how to increase their cost effectiveness by achieving high performance and by process improvements. From the viewpoint of performance, an example with a strength of 34.8 g/d and a modulus of 1420 g/d has been reported and its commercialization is desirable. With respect to production processes, development of practical heat treatment processes is important and requires accumulation of technical experience. On the other hand, there may be another route, i.e. to produce a high-strength as-spun fiber of 20 g/d or more and subject it only to simple heat treatment. This chapter has not dealt with various other properties of thermotropic liquid-crystal polymer fibers, for which some other reviews are available.[31,32]

References

1 Carborundum, US Pat. 3,637,595, 1972.
2 Jackson, Jr W J and Kuhfuss H F, *J. Polym. Sci., Polym. Chem. Ed.*, **14**, 2043, 1976.
3 Calundann G W and Jaffe M, 'Anisotropic Polymers, their synthesis and properties', Proceedings of the Robert A. Welch Conference on Chemical Research, XXVI. Synthetic Polymers, 1982.
4 Inoue M, *Plastics Age*, **140**, May, 1990.
5 Yang D K and Krigbaum W R, *J. Polym. Sci. Part B, Polym. Phys.*, **27**, 1837, 1989.
6 Zimmerman J, *J. Appl. Polym. Sci.*, **39**, 2067, 1990.
7 Acierno D, La Mantia F P and Polizzotti G, *Macromolecules*, **15**, 1456, 1982.
8 Eastman Kodak, Jap. Pat. (Examined), 81-18,016; US Pat. 3,778,410.
9 Zachariades A and Logan J A, *Polym. Eng. Sci.*, **23**, 797, 1983.
10 Mitchell G R and Windle A H, *Polymer*, **23**. 1269, 1982; *Polymer*, **24**, 1513, 1983.
11 Zachariades A, Economy J and Logan J A, *J. Appl. Polym. Sci.*, **27**, 2009, 1982.
12 Unitika, Jap. Pat. (Laid open), 88-264914, 89-272812.
13 Jackson Jr W J, *Brit. Polym. J.*, **12**, 154, 1980.
14 Gutierrez G A, Chivers R A, Blackwell J, Stamatoff J B and Yoon H, *Polymer*, **24**, 937, 1983.
15 Wissbrun K F, Kiss G and Cogswell F N, *Chem. Eng. Comm.*, **53**, 149, 1987.
16 Krigbaum W R, Liu C K and Yang D K, *J. Polym. Sci. Part B, Polym. Phys.*, **26**, 1711, 1988.
17 Yoon H N, *Colloid & Polym. Sci.*, **268**, 230, 1990.
18 Hoechst–Celanese, US Pat., 4,473,682.
19 Hoechst–Celanese, Jap. Pat. (Laid open), 79-77691; US Pat. 4,161,470.

20 Sumitomo Chemical, Jap. Pat. (Laid open), 83-191219; US Pat. 4,503,005.
21 Du Pont, Jap. Pat. (Examined), 83-40976; US Pat. 4,159,365.
22 Du Pont, US Pat., 4,699,748, 1987.
23 Du Pont, Jap. Pat. (Examined), 80-20008; GB Pat. 1,507,207.
24 Carborundum and Dart Industries, Jap. Pat. (Examined), 82-24407.
25 Sugimoto and Asai, *Japan Plastics*, **37**, 103, 1986.
26 Donald A M, Viney C and Windle A H, *Polymer*, **24**, 155, 1983.
27 Uedak, *Sen-i Gakkaishi*, **43**, P-135, 1987.
28 Xiao C F and Takahashi T, *Sen-i Gakkaishi*, **46**, 49, 1990.
29 Yang D K and Krigbaum W R, *J. Polym. Sci. Part B, Polym. Phys.*, **27**, 1837, 1989.
30 Erdemir A B, Johnson D J, Karacan I and Tomka J G, *Polymer*, **29**, 97, 1988.
31 Kanamaru K, *Kogyo Zairyo*, **37**, 36, 1989.
32 Beers D E and Ramirez J E, *J. Text. Inst.*, **81**, 561, 1990.

8

Gel spinning processes

Hiroshi Yasuda, Kaoru Ban and Yasuo Ohta
Toyobo Co. Ltd, Ohtsu, Japan

8.1 Introduction

Although the term 'gel spinning' has recently become familiar through the success of the ultra-high strength polyethylene fiber (UHSPE) development,[1–3], this term has not yet been clearly defined. Here we define gel spinning as a spinning method for high strength fibers through a gel-like state as intermediate substance.

Already, many reviews on the introduction of UHSPE and the related processes have been published.[1–4] In this chapter, gel spinning for UHSPE will be reviewed by focusing on the spinning technologies. In addition, the application of gel spinning to polyvinylalcohol (PVA) and poly(acrylonitrile) (PAN) will be summarized.

8.2 Spinning technology of ultra-high strength polyethylene fiber

The spinning process for UHSPE is schematically shown in Fig. 8.1. All the technical points (Fig. 8.1(b)) are dedicated to reducing macro- and micro-scopic defects to produce an ideal fiber of nearly 100% crystalline structure composed of molecular chains almost perfectly oriented in the fiber direction. The ideal structure and the basic concept of gel spinning are also schematically shown in Fig. 8.2(b). In this chapter, the technical requirements of the gel spinning process will be reviewed from this viewpoint of defect reduction.

8.2.1 Raw materials and solution make-up process

8.2.1.1 Polyethylene

The theoretical and actual strengths and crystalline moduli actually achieved for various kinds of flexible polymers are compared in Table 8.1.

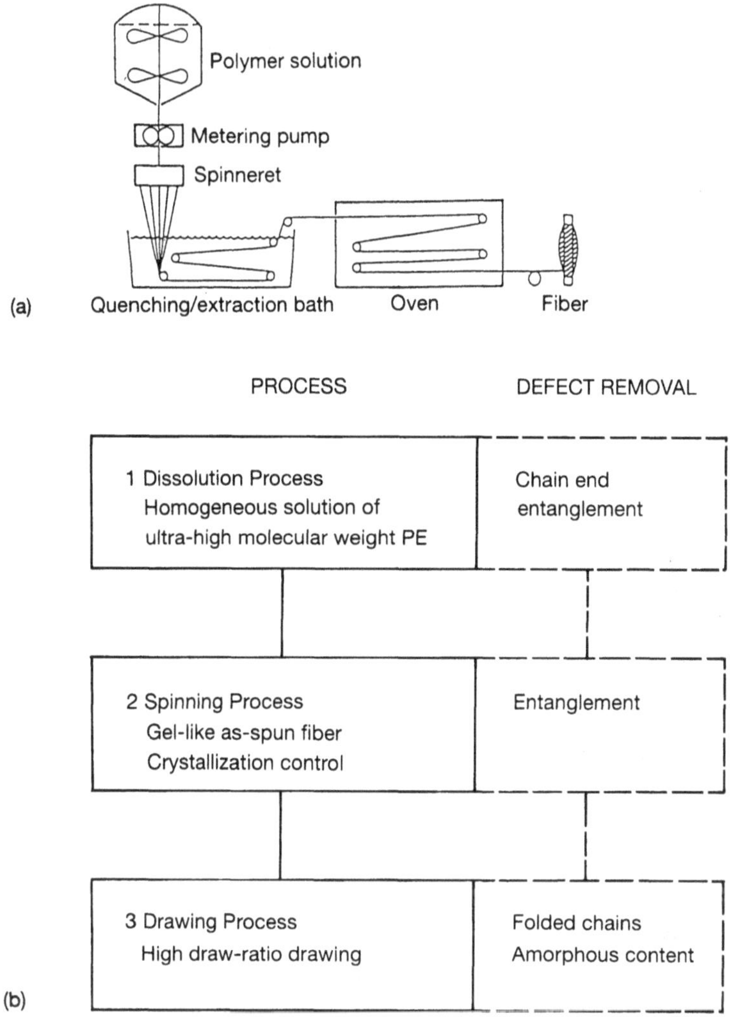

8.1 Schematic diagram of (a) gel spinning process, (b) technical points of gel spinning process.[1]

The reason why gel spinning has succeeded first with polyethylene may be explained by the following properties of that polymer:

1 The highest values of theoretical tenacity and modulus among flexible polymers.
2 A simple planar zig-zag chain structure without any bulky side group.
3 High crystallinity.
4 Lack of high inter-molecular weight polymerization.
5 Availability of ultra-high molecular weight polymerization interaction, like hydrogen bonding.

Table 8.1 Theoretical and achieved tensile properties of high performance fibers

Polymer		Theoretical strength (1) (chain bond)[2]		Theoretical strength (2) (thermo-dynamics)[5]		Actual strength		Theoretical Crystalline Modulus[4]		Fiber modulus	
						Commercialized value	Achieved strength[6]			Commercialized value	Achieved modulus[6]
		g/d	GPa	g/d	GPa	g/d	g/d	g/d	GPa	g/d	g/d
PE	flexible	372	32	80	6.9	9.0		2775	240	100	860
UHMwPE	flexible					30–40	72[7]			1000–2000	2697
Polyamide-6	flexible	316	32	66	6.7	9.5	9.8	1406	142	50	187
PPTA	rigid	235	30	–	–	18–28		1500	183	400-1100	
PET	flexible	232	28	41	5.0	9.5	15.5[8]	1023	125	160	285[8]
WAP	rigid	–	–	–	–	23				560	
POM	flexible	264	33	–	–	11.7		424	53	310	~400
PVA	flexible	236	27	51	5.9	11–17[9]	44[10]	2251	255	230–387[9]	1040[10]
PP	flexible	218	18	32	2.7	9.0	18	423	34	120	489
PAN	flexible	196	20	–	–	5–14	23[11]	833	86	85	268[11]
PVC	flexible	169	21	28	3.6	4.0		–	–	45	

As explained later in detail, of these points, 3 and 5 contribute to reducing defects, and 2 and 4 are important from the viewpoint of drawing performance.

8.2.1.2 Influence of degree of polymerization

The degree of polymerization is the most important property. In gel spinning, ultra-high molecular weight polyethylenes (UHMwPE) having a weight average molecular weight above 600 000 g/mol are used. The reason why such high molecular weight is necessary is believed to be as follows: supposing the perfect orientation of all molecular chains, the chain ends still remain as defects and these defects will act as the starting point of micro-crack formation which causes lower tensile properties.[12,13] According to another argument an ultra-high molecular weight itself does not directly contribute to tenacity, but higher molecular weight is necessary for the drawing, in order to support the drawing force through entanglement points.[5] Indeed, although the molecular weight of PPTA fibers is in the low 10 000s, much lower than that of polyethylene, they still possess high enough strengths and moduli. In the case of para-aramid fibers, the chain orientation is mainly due to the liquid crystalline property derived from their rigid molecular chains, therefore the high molecular weight of PE is not necessary.

The effect of molecular weight on the final properties will be discussed in the next section in conjunction with the effect of concentration.

8.2.1.3 Influence of concentration and control of entanglement

Although successful direct melt spinning of UHMwPE under the critical process window has recently been reported,[14] generally melt spinning of UHMwPE is impossible. These polymers possess a very high number of chain entanglements per molecule, caused by their very long molecular chains, as shown schematically in Fig. 8.2; these long molecular chains cannot pass through each other in the time scale of the drawing

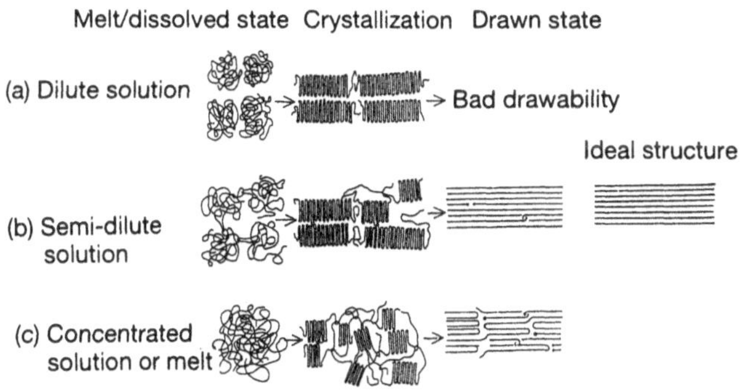

8.2 Schematic explanation of entanglement control concept.[15]

deformation. Therefore, they exhibit extremely high melt (shear) and elongational viscosities, which limit the processability of UHMwPE.

Essential to the formation of high strength fibers is control of this entanglement. In gel spinning, the entanglements are controlled by employing two concepts. One is dilution in an appropriate solvent, and the other is morphological control through the crystallization process.

Various solvents, such as tetralin, decalin, naphthalene, mineral oil, paraffin oil and paraffin wax, have been reported for gel spinning of UHMwPE. From the viewpoint of defect reduction, since entanglement itself behaves as a defect, one entanglement per chain should be enough to transfer the drawing stress between chains. This condition will be achieved at a critical concentration (C^*) at which the random coil chains composed of single molecules begin to overlap each other. C* can be formulated as a function of the average molecular weight (M)

$$C^* \propto M_e/M \quad [8.1]$$

where M_e is the molecular weight between entanglement points.

As expressed in this equation, the ideal (critical) concentration (C^*) becomes lower with increase of M. In other words, this low concentration, like C^*, will permit ideal molecular drawing of UHMwPE up to the theoretical maximum draw ratio. If one postulates the drawing of a single molecule from its random coil state, the maximum draw ratio can be obtained from the following equation[16]

$$\lambda_{MAX} = cM^{0.5} \quad [8.2]$$

where c is a material constant.

As expressed by this equation, the maximum draw ratio increases with the molecular weight (M). If the drawing of UHMwPE gel-like fiber is performed under such ideal concentration conditions, the above argument will be reasonable. However, process economics require that the actual spinning should be conducted at a much higher concentration ($C > C^*$). This region is a 'semi-dilute' or 'concentrate' state in terms of solution rheology. Recently, Bastiaansen has reported on the maximum draw ratio of gel-like fibers as functions of both concentration (C) and molecular weight (M) as shown in Fig. 8.3 and formulated in equation 8.3.[17]

$$\lambda_{MAX} \infty\ cM^{-0.5}C^{-0.5} \quad [8.3]$$

Contrary to equation 8.2, the maximum draw ratio decreases with increase of the molecular weight and concentration. This can be explained by an increase of entanglement number. Therefore, as low a concentration as possible is preferable for higher drawability ($C \sim C^*$). Theoretical and experimental C^* values for ultra-high strength polyethylene having molecular weights above 1 000 000 g/mol have been reported to be 0.5–0.7 wt%.[18] In the actual industrial process, the

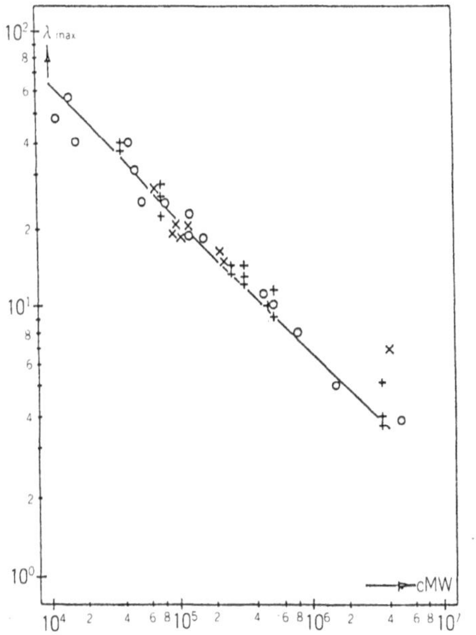

8.3 Relationship between maximum draw ratio, concentration and molecular weight.[17]

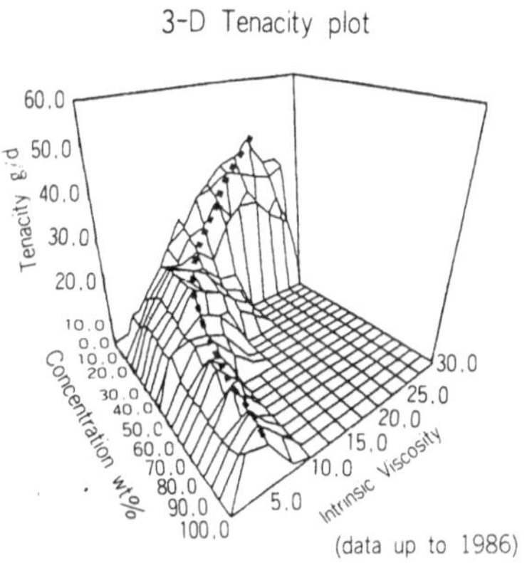

8.4 Relationship between achieved tenacity, concentration and intrinsic viscosity.[19]

optimum concentration and molecular weight are defined as shown in Fig. 8.4.

The above argument assumes that the entanglement points of semi-dilute or concentrated solutions can act as permanent cross-link points even after being solidified. For a lower molecular weight polymer like conventional polyethylene, the higher number of chain ends can easily pass each other through the entanglement point to produce a high level of so-called disentanglement during the drawing process, which makes the orientation of lower molecular weight polymers very difficult, because disentangled chain ends are likely to relax to the random coil state. This is why ultra-high molecular weight is necessary for ultra-drawing of UHSPE.

Even with UHMwPE, the disentanglement possibly occurs during the spinning process due to crystallization, which reduces the entanglement number to some extent. This mechanism allows production of UHSPE even at much higher concentration ($C > C^*$) as shown in Fig. 8.4. This will be discussed again in Section 8.2.2.

Furthermore, although equation 8.3 was obtained under ideal drawing conditions, i.e. with an extremely low drawing speed, in actual industrial processes, which require much higher drawing speeds, the dominant factor for the maximum draw ratio will be quite different. In high-speed drawing processes, the viscoelastic properties become more dominant, i.e. the extremely long relaxation time of UHMwPE affects drawing performance. This point will be discussed again in Section 8.2.3.

8.2.1.4 Dissolution process

A relatively high-concentration polyethylene solution must be prepared as uniformly as possible, because any inhomogeneity will remain as a defect in the final fiber structure and reduce the final mechanical properties of the fibers.

In Fig. 8.5, the shear viscosity change obtained with a 5wt% decalin solution of UHMwPE having $Mw \sim 2\,000\,000$g/mol is shown as a function of temperature. Near the melting point, a remarkable viscosity rise and the so-called Weissenberg effect are observed, and both result in inhomogeneity of solution concentration. Furthermore, once the dissolved polymer forms a gel-like solution, the heat transfer coefficient of that gel-like solution decreases remarkably, and this also causes an inhomogenous temperature distribution locally within the solution.

To avoid this problem, various dissolving methods have been reported, including the use of a screw-type extruder, pouring hot solvent on to partially swollen polymer, and forming gel-spherulites.[20,21] Details can be found in the patents.

8.2.2 Spinning process

8.2.2.1 Polymer flow

From the viewpoint of the spinning process, understanding the rheological properties of UHMwPE solution is of first importance. However, due to the difficulty of UHMwPE polymer characterization,

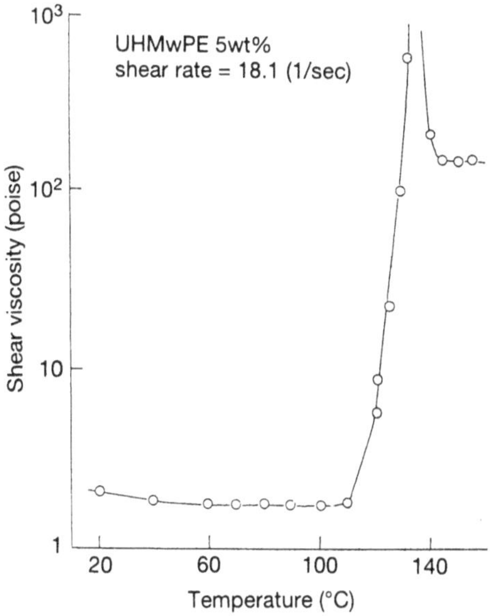

8.5 Viscosity change of UHMwPE/decalin mixture under conditions of continual temperature rise.

not many articles have been published on the rheological properties of UHMwPE solution.

The fundamental rheological constants of UHMwPE solution obtained by a capillary type viscometer are compared with those for melt spinning conventional PE and PET, under typical spinning conditions for each polymer, in Table 8.2. In this rough comparison, the characteristic feature of UHMwPE solution can be summarized as its rubber elastic property; UHMwPE solution shows a combination of lower shear viscosity and shear modulus than expected from its molecular weight, and shows a much higher relaxation time, conventionally obtained from the ratio of shear modulus (G) and shear viscosity (η). This higher characteristic relaxation time causes a so-called memory effect of UHMwPE solution which means the memory of the stress applied earlier in the process at the inlet part of the spinneret hole, which affects the rheological response later in the process. Therefore, carcful design of the polymer flow, especially in the spinneret, is required.

In Fig. 8.6, the shear rate dependence of the shear viscosity of UHMwPE solution is compared with the case of PET melt spinning.[22] As shown in this figure, over a wide range of applied shear rate this solution shows unique non-Newtonian behavior. This solution property causes inhomogeneous distribution of viscosity at each local flow point. Again, careful design of the spin-line, especially the spinneret, is necessary. As an extreme example, the dimensional change of extruded UHMwPE

Table 8.2 Characteristic rheology constants for solution and melt spinning

	UHMwPE5% (Solution) Mw ~ 2000kg/mol $T_{spin} = 150°C$	Conventional PE (Melt) Mw ~ 180kg/mol $T_{spin} = 180°C$	PET (Melt) M_w ~ 80kg/mol $T_{spin} = 280°C$
Shear rate (1/sec)	1000	~630	1000
Shear modulus (dyne/cm^2)	$\sim 9 \times 10^3$	$\sim 5 \times 10^5$	$\sim 5 \times 10^5$
Relaxation time (sec)	$\sim 17 \times 10^{-3}$	$\sim 10 \times 10^{-3}$	$\sim 2 \times 10^{-3}$
Shear viscosity (poise)	~150	~5000	~1000

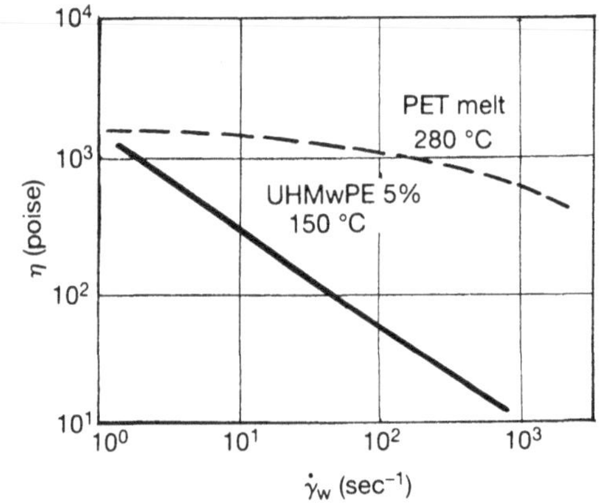

8.6 Shear rate dependence of shear viscosity of UHMwPE solution in comparison with PET melt.

solution as a function of spinning speed is schematically demonstrated in Fig. 8.7. At lower spinning speed, the extruded solution exhibits quite large die-swell, which is related to the highly elastic property of UHMwPE solution. With increasing spinning speed, the size of the die-swell become smaller due to the stretching under the spinneret, and at much higher spinning speed, a so-called 'pull-out' phenomena can be observed up to filament breakage, where the solution completely loses its die-swell and may even peel off from the inside wall of the spinneret capillary. This typical feature may also explain the combination of higher elongational strength of the solution and its highly elastic properties.[23]

From the above discussion, it can be concluded that the characteristic feature of UHMwPE solution is its non-Newtonian and highly elastic property. One of the main technical points of UHMwPE gel spinning is the flow management of such a highly elastic solution.

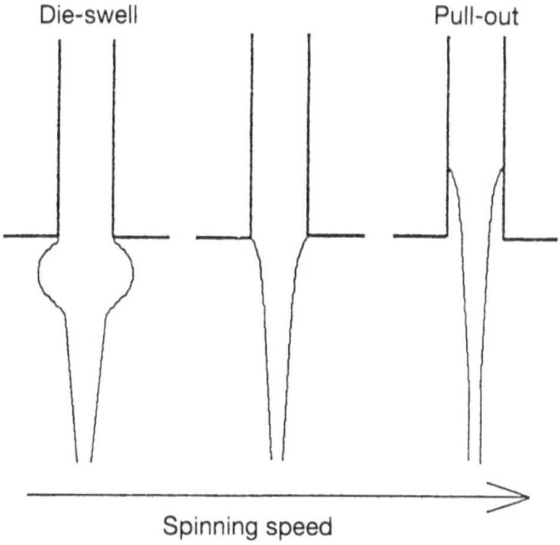

8.7 Schematic explanation of extrudate behaviour in terms of spinning speed.

8.2.2.2 Crystallization

In the gel spinning process, the extruded solution is substantially cooled down by a gas or a liquid cooling medium, and so crystallization occurs. During this crystallization process, some part of the entanglement is considered to be lost because the entanglement points cannot be incorporated in the crystal and the chain will be disentangled before crystallization. Hence, disentanglement due to crystallization can be anticipated and this is another technical point for the success of gel spinning even at high concentration.

Recently, Pennings *et al.* reported the high speed spinning of UHMwPE solution, and a remarkable tensile strength value of 26g/d at a spinning speed of 1000m/min without further drawing. This success implies that more effective disentanglement might happen during the spinning process, i.e. disentanglement due to spinning stretch.[24] An investigation of the structure development mechanism during crystallization of UHMwPE solution may be worthwhile.

Through the crystallization process, the solution is solidified into a more rigid gel-like structure having dispersed crystallites connected by a small number of entanglements remaining as pseudo-crosslinking points. Such a structure is ideal for the drawing process as explained in Fig. 8.2.

8.2.3 Drawing process

As mentioned in the previous sections, the drawing performance of a gel-like as-spun filament is influenced by the dissolving conditions (concentration, molecular weight etc.) and also by the morphological structure of the gel-like fiber. Moreover, those influences seem to be unified by the entanglement concept as expressed in equation 8.3.

If an actual drawing process is performed with an ideally low drawing speed, the effective and homogeneous deformation of micro-structure leads to ideal drawing as expressed in equation 8.3.

On the other hand, another important requirement for a practical drawing process is a higher drawing speed with minimum yarn breakage. To fulfil this requirement, the following relationship for the molecular deformation should hold.[1]

$$\lambda_{MAX} \propto \nu \sim 1/\tau \qquad [8.4]$$

where ν is the deformation rate and τ is the characteristic relaxation time. If one applies higher deformation rate ($\nu > 1 / \tau$) the molecular chain cannot relax any extra localised stress, so molecular chain breakage occurs. On the other hand, at $\nu < 1 / \tau$, the molecular chain has time to relax to some extent, which makes disentanglement possible.

The value of τ is affected by the molecular structure of the as-spun fiber structure, specifically by chain entanglement, and is therefore strongly dependent on both the molecular weight and the concentration. From the molecular theories which deal with the dependence of the relaxation time on the molecular weight and concentration, τ can be generally expressed as follows:

$$\tau \propto C^{\alpha} M^{\beta} \qquad [8.5]$$

and from equation 8.4

$$\lambda_{MAX} \propto C^{-\alpha} M^{-\beta} \qquad [8.6]$$

Values of these coefficients reported for the Graessley model[15] are $\alpha = 1.5$ and $\beta = 3.5$, and for the Doi–Edwards model[25], $\alpha = 1.0$ and $\beta = 3.0$. These values are much higher than those of equation 8.3, and we have obtained a similar result to these theoretical power laws in our high-speed drawing experiments.

Since the solvent is removed either fully or partly during the spinning and drawing processes, the above ideals cannot yet be completely accepted. However, it can be concluded at least that the entanglement density and/or entanglement structure, which are mainly determined through the solution make-up and spinning process as described in the previous sections, dominate the drawing performance of UHSPE.

Another important factor which dominates the drawing performance is the ease of pulling out molecular chains from the crystalline structure. This is particularly easy with polyethylene, because it has no strong interchain interactions like hydrogen bonding. On the other hand, this also causes lower creep resistance of UHSPE fibers.

8.2.4 Features of ultra-high strength polyethylene fiber and future development

8.2.4.1 Features and use of Dyneema SK60™

The performance of Dyneema SK60 as a representative UHSPE fiber is summarized in Table 8.3. Many applications are making progress, notably high performance ropes, high performance fabrics, and reinforcements for composites. These applications utilize UHSPE fibers' excellent properties such as light weight, super-high strength and modulus, good impact properties, environmental and chemical stability.

8.2.4.2 Future development

Competition among high performance fibers will prompt further improvement in UHSPE performance and productivity. In particular, strength and modulus will be improved towards the theoretical values. For example, experimentally, a strength of 72g/d which is close to a theoretical value[5] has been reported.[7] Development will focus on how to realize such laboratory scale trials as actual processes at higher drawing speeds with low yarn breakage rates. Besides the tensile properties, deficiencies like lower heat resistance, poor adhesion and low creep resistance will also be improved.[26, 27] Improvements in productivity will also be brought about. For example, the ultimate no-yarn-breakage process and new technologies such as high-speed spinning, reported by Pennings *et al.*,[24] are worth developing.

8.3 Gel spinning – other flexible polymers

The success of the gel spinning process for UHSPE has prompted the application of this technology to other flexible polymers. Here the application to two representative flexible polymers, polyvinylalcohol (PVA) and polyacrylonitrile (PAN), is introduced.

8.3.1 PVA

PVA polymer is the most promising candidate for the next gel spinning application. As shown in Table 8.1, the theoretical strength of PVA is 236g/d and its crystalline modulus is 2251g/d. These values are close to those of PE, therefore higher strength and modulus can be expected. In fact, recently a strength as high as that of aramid fibers has been realized. Generally speaking, the fundamental concept for gel spinning of PVA is similar to that of PE as described in this article. A characteristic feature of PVA is that a major part of the effort has gone into control of the interchain hydrogen bonding.[4,9,28]

Many attempts to produce high strength PVA fibers have been reported in patents and articles. For example, in patents[29] solvents such as glycerol, ethylene glycol and water are used for PVA having a degree

Table 8.3 Comparison of Dyneema SK60 with other high performance fibers

		Dyneema SK60	Aramid fiber	Carbon fiber		E-glass fiber	Polyester, polyamide HT filament	Steel fiber
				HT type	HM type			
Density	(g/cm$^{3)}$	0.98	1.4–1.5	1.7–1.9	1.7–1.9	2.54	1.1–1.4	8
Tensile strength	(g/d) (kg/mm^2)	30–45 260–400	22 290	17–22 280–350	12–15 200–250	9.6 220	9–10 100-110	4 280
Modulus	(g/d) (kg/mm$^{2)}$	1000–1400 8800–13 000	1500–1000 6000–13 000	1200–1500 20 000–25 000	2000-2500 35 000–40 000	300 7000	50–100 500–1200	260 19 000
Elongation to break	(%)	2–5	2–4	1.0–1.5	0.5	4.0	13–19	2
Thermal behavior			Melt 150°C 420°C	Degraded	2500°C	2500°C 730°C	Melt 240°–260°C	Melt

of polymerization more than 2000. The gel-fibers obtained by cooling these solutions are extracted by alcohol and then stretched more than 20-fold.

Achieved strengths of 22g/d (by adding boric acid[30]), 24 g/d (by using as solvent DMSO/H_2O = 80/20[31] and drawing at a draw ratio of 48), and again 24 g/d (by using high molecular weight PVA having a degree of polymerization above 10 000 and applying multi-stage drawing[32,33]) have been reported. Furthermore, an extremely high strength of 44g/d and a modulus of 1040g/d were obtained by gel-spinning using highly syndiotactic PVA having a high degree of polymerization of 15 000 and glycerol as solvent.[10] This report suggests that the tacticity of PVA is also important. Of course, higher molecular weight causes an increase in the tensile strength, but higher draw ratio sometimes results in lower properties.[9] This may be explained by the pressure of strong interchain bonds.

8.3.2 PAN

Representative information from patents[11] shows that a tenacity around 20g/d has been reported by dissolving PAN having a molecular weight more than 1 000 000 in NaSCN/H2O to form a 5–10% solution, coagulating in a lower temperature bath to form gel-like fiber, and drawing first in hot water or glycerol then in the dry state. Commercially, tenacities around 13–15g/d have been announced.

8.4 Conclusion

The industrial success of the development of UHSPE has stimulated the application of 'gel technology' to other organic polymers. Some successful results have been reported with PVA and PAN, which shows that gel spinning technology is a concept of general utility as a processing route to high-performance fibers using flexible polymers. However, even with UHSPE, the mechanism of structural development during spinning and drawing is not well understood. Progress in such fundamental understanding should lead to further advances of this technology in future.

References

1 Lemstra P J, Kirshbaum R, Ohta T and Yasuda H, in *Developments in Oriented Polymers–2*, Ed I M Ward, Elsevier Applied Science, p. 39 1987.
2 Ohta T, *Sen-i Gakkaishi*, **40**, P-407, 1984.
3 Ohta Y and Sugiyhama H, *Polymer Application*, **38**, 68, 1989.
4 Kunugi T, Ohta T and Yabuki K, *Ko-Kyodo Ko-Danseiritsu Sen-i (High-Strength High-Tenacity Fibers)*, Kyo-ritsu Shuppan, Tokyo, p. 58, 1988.

5 Smith Jr K J, *Polym. Eng. Sci.*, **30**, 437, 1990.
6 Miyasaka K, *Sen-i Gakkaishi*, **46**, P-431, 1990.
7 Matsuo M, Inoue K and Abumiya N, *Sen-i Gakkaishi*, **40**, T-275, 1984.
8 Murase Y and Nagai A, Collection of Summer Seminar Preprints of the Society of Fibre Science and Technology, Japan, P-13, 1990.
9 Narukawa H and Noguchi H, *Sen-i Gakkaishi*, **46**, P-466, 1990.
10 Yamaura K, Tanigami T and Matsuzawa S, *Polym. Prep. Japan*, **38**, 4400, 1990.
11 Mitsubishi Rayon, Jap. Pat. (Laid Open) 89-104816; Japan Exlan and Toyo Boseki, Jap. Pat. (Laid Open) 86-97415; US Pat. 4,658,004; Toray, Jap. Pat. (Laid Open) 88-182317.
12 Termonia Y, Meakin P and Smith P, *Macromolecules*, **18**, 2246, 1985.
13 Smook J, Hamersma W and Pennings A J, *J. Mater. Sci.*, **19**, 1359, 1984.
14 Waddon A J and Keller A, *J. Polym. Sci. Polym. Phys. Edn.*, **28**, 1063, 1990.
15 Graessley W W, *J. Chem. Phys.*, **54**, 5143, 1971.
16 Smith P, Lemstra P J and Booij H C, *J. Polym. Sci.*, **19**, 877, 1981.
17 Bastiaansen C W M, *J. Polym. Sci, Polym. Phys Edn.*, **28**, 1475, 1990.
18 Matsuo M, *Nihon Reoroji Gakkaishi*, **13**, 4, 1985.
19 Kirshbaum R, Yasuda H and van der Gorp E H M, *Industrie-Textilien*, **36**, 134, 1986.
20 Academy of Ind. Sci. Tech. Jap. Pat. (Laid open), 84-78238.
21 Toyobo Jap. Pat. (Laid open), 85-45630.
22 Yasuda H, Ohta Y and Ban K, *Sen-i Gakkaishi*, **47**, P-595, 1991.
23 Bulters M J H and Meijer H E H, *J. Non-Newtonian and Fluid Mech.*, **38**, 43, 1990.
24 Pennings A J, Roukema M and van der Veen A, *Polym. Bulletin*, **23**, 353 1990.
25 Doi M and Edwards S F, *The Theory of Polymer Dynamics*, Oxford University Press, 1986.
26 Toyobo Jap. Pat. (Laid open), 88-175111.
27 Dyneema, Bastiaansen C W M and Toyobo, US Pat. 5,128,415.
28 Shibayama M, *Kino Zairyo*, **3**, 16 , 1988.
29 Jap. Pat. (Laid Open) 85-108711; Allied Corp., Jap. Pat. (Laid Open) 84-130314, US Pat. 4,490,711; Kuraray, Jap. Pat. (Laid Open) 87-85013; Stamicarbon, Jap. Pat. (Laid Open) 87-90308, 87-90309, US Pat. 4,810,450, 4,812,277.
30 Fujiwara H, Shibayama M, Chen J H and Nomura S, *J. Appl. Polym. Sci.*, **37**, 1403, 1989.
31 Ikada Y *et. al.*, *Polym. Prep. Japan*, **38**, 4406, 1989.
32 Kunugi T, *Polymer Application*, **38**, 484, 1989.
33 Kunugi T, Kawasumi T and Ito T, *J. Appl. Polym. Sci.*, **40**, 2101, 1990.

9

Spinning of ultra-fine fibers

Miyoshi Okamoto
Toray Industries Inc., Ohtsu, Japan

9.1 Introduction

Although there is no exact specification, an ultra-fine fiber is conventionally defined as a fiber of less than 0.7 denier. Ultra-fine fibers can be extruded by reducing the polymer output at the spinneret and drawing with a large draw ratio.[1–11] However, polyester for instance, cannot be extruded at less than about 0.15 g/min, because the monofilament will break during the fiber-forming process just after the extrusion. No drawn ultra-fine fiber less than 0.3 denier has been obtained by conventional extrusion.[1,12]

Since no application of such ultra-fine fibers was foreseen in the 1960s, there had been no technical or commercial interest in them until Toray put the new suede-like material Ecsaine® (Ultrasuede® in USA; Alcantara® in Europe)[3] on the market in 1970 (Fig. 9.1). As seen from Fig. 9.1, ultra-fine polyester fiber can reproduce deer suede from the microscopic as well as from the macroscopic structural point of view. Ecsaine® is made of homogeneous ultra-fine polyester fibers of less than 0.1 denier and marked the start of a new epoch in the appearance and handle of clothing materials. Ultra-fine fibers are also applied to the

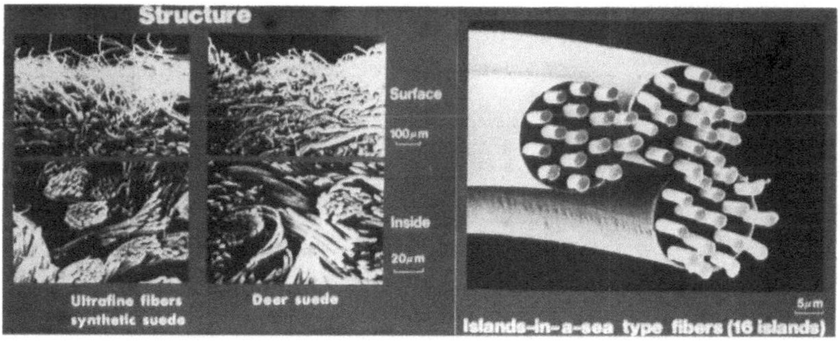

9.1 (a) Microscopic structure of synthetic suede of ultra-fine fibers (left) and deer suede (right); (b) Islands-in-a-sea type fiber showing location of ultra-fine fibers.

fabrication of silk-like materials and have revealed characteristics surpassing natural silk. As a result, ultra-fine fibers of 0.1 to 0.5 denier have attracted much attention in Europe and the USA.

Ultra-fine fibers were produced by melt-blow spinning and flash spinning in the late 1950s.[1–5]. These fibers were not of the continuous-filament type but were fine staple fibers of random length which found no application except for being processed into nonwoven sheets immediately after spinning.

Ultra-fine fibers of a continuous-filament type have a relatively recent history. A petal-shaped conjugate fiber described in a Du Pont patent[13] was probably the first example of a potential ultra-fine filament. This patent was issued in 1961 as one of several patents for the production of fibers with a triangular cross-section. It was aimed at producing a fiber with a sharp edge by utilizing the boundary of two components *A* and *B*. Another patent, issued simultaneously from Du Pont,[14] described splitting two-component conjugate fibers of non-circular cross-section into the two separate components after weaving. No attention was paid at that time to combining these technologies to produce ultra-fine fibers, since the fiber with a sharp edge was the primary concern.

Okamoto *et al.* of Toray developed conjugate spinning technology for the production of ultra-fine filaments in the mid 1960s by increasing the splitting number of *A* and *B* components. Here two components, *A* and *B*, are arranged alternately and extruded to yield a conjugate filament which is split into ultra-fine fibers after processing. This was the first attempt to produce an ultra-fine fiber intentionally. Matsui *et al.* of Kanebo also tried multi-layer conjugate spinning in 1968 for the production of flat ultra-fine filaments.[15]

9.2 Outline of ultra-fine fiber products and processes

9.2.1 Definition and types of ultra-fine fiber

The definition of ultra-fine fiber has varied according to the convention employed, for example, in articles published in newspapers and magazines.[9–11,16–18] For instance, the Textile Committee, Germany, defines a micro-denier fiber as a fiber finer than 1.2 dtex for polyester and finer than 1.0 dtex for polyamide. Although a rather thick (1 denier or more) fiber is sometimes claimed as an ultra-fine fiber commercially, an ultra-fine fiber should preferably be specified as a fiber of less than 0.5 d. This chapter therefore deals with filaments of approximately 0.05–0.5 d. A fiber of less than 0.1 d is sometimes referred to as a super ultra-fine fiber.

Ultra-fine fibers are classified into two types: (i) a continuous-filament type and (ii) a random (staple) type. Since recent developments in the field of ultra-fine fibers have focused on the continuous-filament type, as exemplified particularly by suede-like artificial leather or shingosen, this chapter will concentrate on this type.

9.2.2 Manufacturing processes for ultra-fine fibers [1–34]

9.2.2.1 Continuous-filament type

Ultra-fine fiber of the continuous-filament type is now produced by a variety of methods including:

1 Direct spinning (conventional extrusion).
2 Conjugate spinning (extrusion of polymer components arranged alternately):
   (a) islands-in-a-sea type;
   (b) separation type or splitting type;
   (c) multi-layer type.

Various processes for ultra-fine filament production are illustrated in Fig. 9.2, where the upper part shows the fibers just after extrusion and the lower part shows them after conversion into ultra-fine fibers.

9.2.2.2 Random (staple) type

Ultra-fine fibers of the random type are produced by:

1 Melt-blowing or jet spinning.
2 Flash-spinning.
3 Polymer-blend spinning.
4 Centrifugal spinning.
5 Fibrillation or violent flexing.

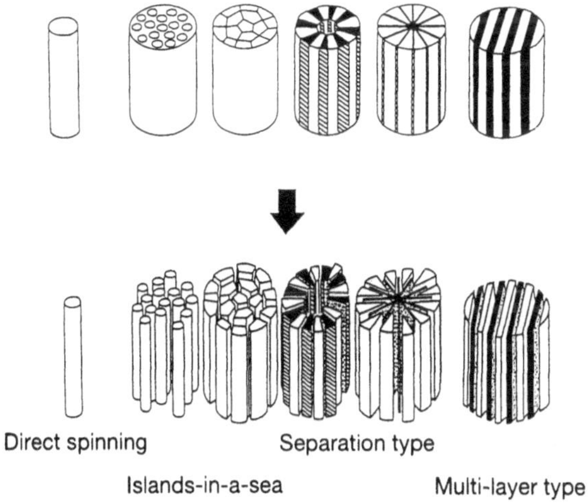

9.2 Spinning of ultra-fine fibers of continuous-filament type. The upper part shows the fibers just after extrusion and the lower part after splitting into ultra-fine fibers.

6 Turbulent flow-moulding.
7 Bursting.
8 Other methods.

9.2.3 Characteristics of ultra-fine fibers

Different ultra-fine fibers are designed to provide the following characteristics:

1 Softness, flexibility and smoothness.
2 Fine textile structure.
3 Micro-pockets in fabrics.
4 High filament density at the textile surface.
5 Large surface area per unit weight, and a characteristic interfacial property.
6 Small radius of curvature (resulting in luster and characteristic color)
7 Large aspect ratio (the ratio of length to diameter) and easy entanglement.
8 Good interpenetrating capacity in other materials.
9 Quick stress relief.
10 Low resistance to bending.
11 Bio-singularity relative to living tissues and fluids.
12 Fine, sharp edges.

Since the cross-sectional area A, the second moment of the cross-section M and the torsion I_p are given by $A = (\pi/4)D^2$, $M = (\pi/64)D^4$, and $I_p = (\pi/32)D^4$, with D being the diameter of the cross-section of a filament, these values decrease exponentially as the diameter D decreases. Thus the flexibility (item 1 in the above list) of ultra-fine fibers is the result of a small cross-sectional diameter.

9.3 Spinning of the continuous-filament type

9.3.1 Direct spinning

The direct spinning method is an extension of conventional spinning, where the spinning conditions are optimized so as to be suitable for the production of ultra-fine fibers. In the application of conventional melt spinning, the following problems can be foreseen:

1 Fiber break-down (dripping).
2 Variation of filament thickness.
3 Spinneret clogging.
4 Denier variability among filaments in a single yarn.

The following precautions are taken in order to avoid these problems:[33]

1 Optimization of polymer viscosity (i.e. a higher spinning temperature to reduce viscosity).
2 Optimization of the spinneret design (i.e. the spinning holes arranged so as to ensure homogeneous cooling).
3 Optimization of the ambient temperature underneath the spinneret (i.e. quenching, cooling rate control).
4 Optimization of filament assembly (i.e. assembly nearer to the spinneret).
5 Optimization of spinning draft (i.e. spinning tension control).
6 Lower rates of extrusion (i.e. stable polymer transmission)
7 Purification of spinning polymer (i.e. high-efficiency filtration).

The Unitika Co. was the first company to put ultra-fine fibers of the order 0.3–0.5 denier on to the market, although Asahi Chemical Industry Co. developed finer polyester fiber of 0.1–0.3 denier by optimizing the polymer melt viscosity, the spinneret design, the ambient temperature underneath the spinneret (i.e. the cooling conditions) and the filament assembly conditions. More precisely, the conditions for production of ultra-fine fibers of less than 0.3 denier are that the polymer melt viscosity should be adjusted to be less than 950 poise, the cross-sectional area per spinneret hole should be less than 3.5×10^{-4} cm^2, the ambient temperature at 1–3 cm underneath the spinneret should be kept below 200°C, and the extruded filaments should be assembled at 10–200 cm underneath the spinneret. The procedure is illustrated schematically in Fig. 9.3.

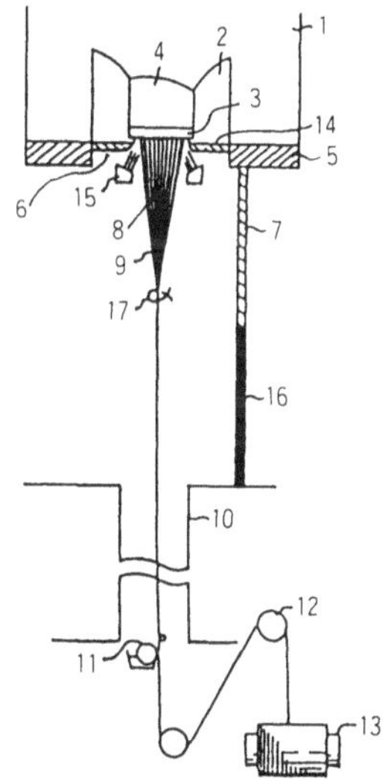

9.3 Schematic diagram of direct spinning: 1. spinning head; 2. spinneret pack; 3. spinneret; 4. polymer melt; 5. insulation for spinning head; 6. thermal insulation plate; 7. blower; 8. thermal annealing zone; 9. spinline; 10. spinning column; 11. oiling roller; 12. forwarding roller; 13. wind-up; 14. thermal insulation plate; 15. blower; 16. draught exclusion plate; 17. convergence guide.

Asahi Chemical Industry Co. has succeeded in producing ultra-fine polyester fiber of less than 0.15 denier by extruding polyester of melt viscosity less than 480 poise through a spinneret with over 300 holes, of less than 1 x 10^{-4} cm^2 cross-sectional area per hole, arranged concentrically. However, the extruded polymer tends to form droplets and exhibits no drawability unless the thermal environment immediately below the spinneret is suitably controlled. The ambient temperature at 1–3 cm underneath the spinneret holes must be kept below 150°C by blowing cold air from the circumference of the spinning threadline to enable the polymer to be drawn into filaments. These concentrically arranged filaments should all be cooled at the same rate. Then the filaments are assembled at 20–70 cm underneath the spinneret holes, and wound up as undrawn fiber. This undrawn fiber can be drawn conventionally to yield ultra-fine fiber of less than 0.15 denier. Table 9.1 summarizes the spinning conditions for ultra-fine polyester fiber production by direct spinning.[33]

The Teijin Co. has investigated the influence of air friction on the high-speed spinning of ultra-fine fiber.[19] Toray and Toyobo have also produced ultra-fine fiber by direct spinning. Unitika has succeeded in producing cationic-dyeable ultra-fine polyester filaments by using homogeneous cooling temperatures at spinning.

Although the PET filament is drawn only 4–6-fold in a conventional process, it can be drawn 10–20-fold in a particular condition called super-draw. Du Pont has proposed a method for producing ultra-fine fibers by super-draw,[1–3] but no industrial application has been implemented so far because of the unstable and restricted conditions required.

Ultra-fine fibers can be produced by wet spinning as shown below. The following points should be observed in the technical application:

Table 9.1 Spinning conditions for PET ultra-fine fiber by direct spinning

	Spinning conditions	
	<0.3d	<0.15d
Number of holes in spinneret	>140	>150
Cross-section of output hole (cm)	$<3.5\times10^{-4}$	$<1.5\times10^{-4}$
Polymer melt viscosity (poise)	<950	<300
Ambient temperature at a point 1–3 cm below spinneret (°C)	<200	<150
Filament-assembling position (cm below spinneret)	10–20	20–70
Drawing	conventional	conventional
Tenacity (g/d)	3–5	3–5
Elongation (%)	20–40	20–40

1 The solvent should be chosen so as not to form a porous structure in the resulting fiber, and the solution concentration (and hence the viscosity) should be adjusted to the optimum level for ultra-fine fiber spinning.
2 The spinning solution should be filtered through a fine filter to eliminate particles larger than 1/3 of the nozzle diameter.
3 The nozzle diameter should be less than 30 μm and a multi-hole system is desirable for industrial purposes.
4 Drawing can be performed by a conventional technique.

In the case of acrylic fiber spinning using a polyacrylonitrile/vinyl acetate/sodium methacryl-sulphonate (92.5/7.0/0.5) copolymer, an ultra-fine acrylic fiber of 0.06–0.4 denier with a tenacity of 3.0 g/denier and 26% extension is produced under the following spinning conditions[33] (Fig. 9.4):

1 Specific viscosity: 0.17–0.19.
2 Solvent: dimethylacetamide or dimethylformamide.
3 Concentration: 16–19% solute.
4 Filter: sintered metal filter of less than 10 μm filtration cut-off.
5 Nozzle diameter: 20–30 μm.
6 Number of nozzle holes: 40 000–80 000.

A single-component ultra-fine fiber is obtained by direct spinning, and the later processes require no complicated processing such as splitting into two components or removing a second component. However, filament breakage and fluff formation are often observed in direct-spinning of ultra-fine fibers, and a handle of high quality cannot be expected.

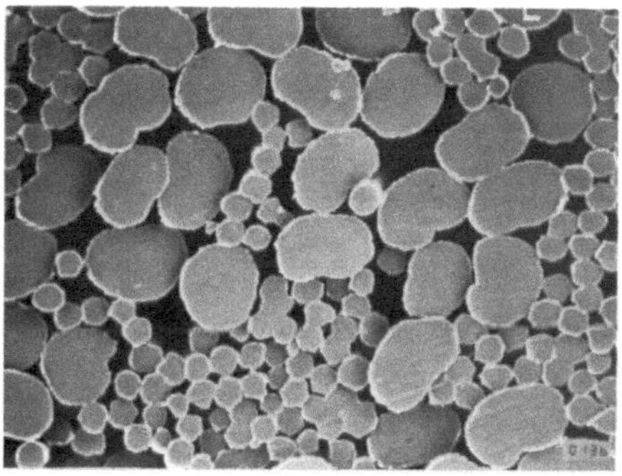

9.4 Cross-section of ultra-fine acrylic fiber mixture (0.1d + 1.5d; Mitsubishi Rayon)

9.3.2 Conjugate spinning with alternately arranged polymers[1,20–22]

The technical problems in direct spinning can be solved by conjugate spinning, which yields homogeneous ultra-fine fibers. The idea of conjugate spinning predates direct spinning for the production of ultra-fine fibers. Okamoto *et al.* (Toray) and Matsui *et al.* (Kanebo) investigated the extrusion of conjugate fibers with a cross-section consisting of highly-dispersed conjugate components by modifying the spinneret structure. Okamoto suggested that the resulting conjugate fibers should be termed 'alternately-arranged-polymer fibers' in order to express correctly both the morphology in the longitudinal direction and the islands-in-a-sea structure in the lateral direction.[1–8,20,21]

Alternately-arranged polymer spinning is classified into two types from the technical viewpoint: (i) the islands-in-a-sea type, where the sea component is removed by dissolving it in a solvent, and (ii) the separation type or splitting type, where Breen of Du Pont's patent[13] has been applied to ultra-fine fiberization. In either case, no drawing is required at spinning and the micro-fiberization is performed in the form of fabrics. No special technical problems arise at later processing, compared with conventional spinning.

9.3.2.1 Islands-in-a-sea or alternately-arranged-polymer spinning

The extrusion principle of the islands-in-a-sea type is shown in Fig. 9.5(a). Here, two-component polymer flows of a conjugate type or a sheath-and-core type are assembled into a single flow, which is extruded from the spinneret shown in Fig. 9.5(b). The number of two-component polymer flows in a bundle determines the thickness of the resultant filaments. Although it depends on the assembling process of two-component polymer flow, the sheath-and-core type has an advantage with respect to obtaining the desired number of split filaments since the components in the splitting type may mingle during the assembling process.

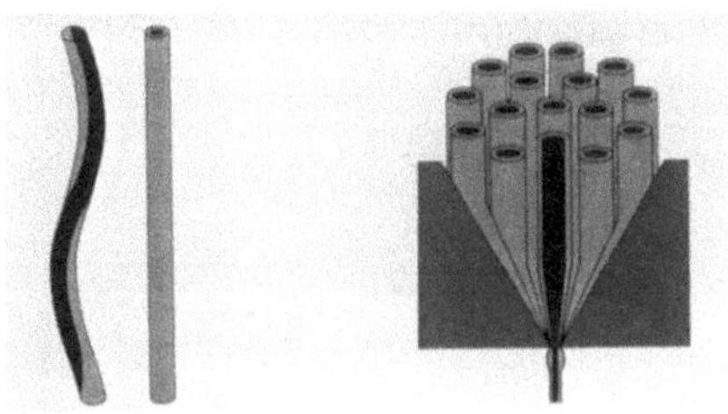

(a) (b)

9.5 Concept of islands-in-a-sea fibers.

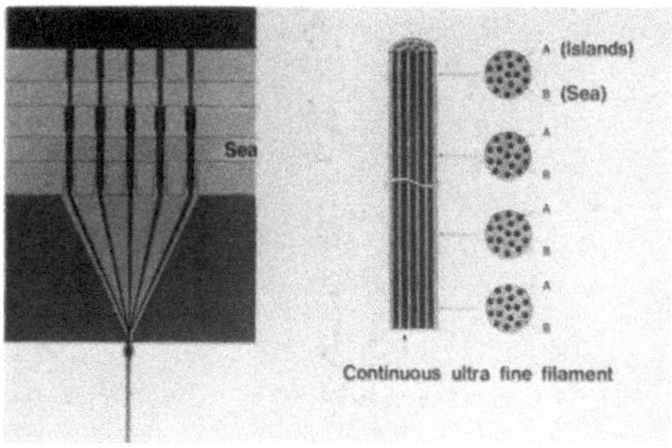

9.6 Islands-in-a-sea fiber: spinneret assembly (left) and fiber structure (right).

Polyester, nylon, polypropylene, polyethylene and polyphenylene sulfide are among the polymers employed as island components. A sea-component, such as polystyrene, a 2-ethylhexyl acrylate copolymer or a copolymer of ethylene terephthalate and sodium sulfo-isophthalate, is removed by dissolving it in a solvent (which must be a non-solvent for the island component) after conventional processing into woven, knitted or nonwoven fabrics. Thus there is no fundamental difference in spinning and further processing from conventional melt spinning. For example, the extrusion temperature is 275–300°C when PET is employed as an island component. The micro-fiberization takes place after the macro-filaments have been processed conventionally into fabrics (see Fig. 9.6).

This technology has provided a means of industrial production of suede-type artificial leather, silk-like fabrics, wiping cloths and fine filters made of ultra-fine fibers. The number of islands in the ultra-fine multifilament yarn is specified by the design of the spinneret. The ratio of the island component to the sea component is determined by the extrusion rate of each component. An example of this procedure is shown in Fig. 9.7, which demonstrates an island-in-a-sea type fiber with 145 islands. Here, the island component is sheathed in the sea component. The spun and drawn fibers are processed into fabrics and micro-fiberized by removing the sea component (Fig. 9.8). Three-component spinning can be carried out with two island-components (for example, where island 1 is PET and island 2 nylon) by designing a three-component spinneret assembly. The sea-component can be reduced to 2–10% of the total components from the purely technological point of view, but the space between the ultra-fine filaments is also reduced and this may lead to poorer handle of the products. When the sea-component is small in amount and not miscible with the island-component, the splitting can be carried out mechanically.

Since the ultra-fine filaments (the island component) are sheathed by the sea-component in the islands-in-a-sea type shown in Fig. 9.9, they are protected from damage during later processing. This technology is

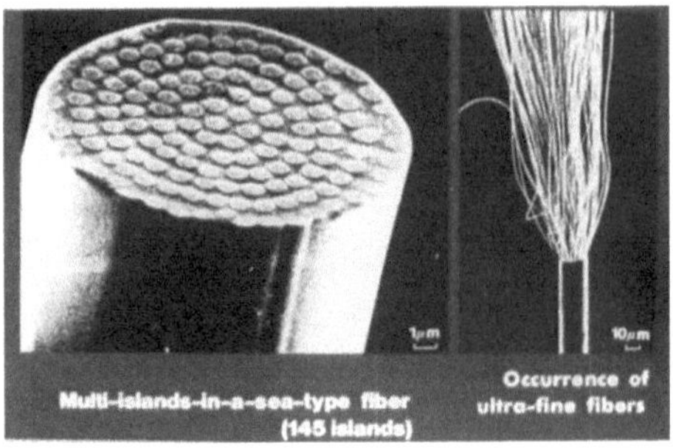

9.7 Islands-in-a-sea fiber with 145 islands (left) and ultra-fine filament yarn obtained by dissolving the sea component (right).

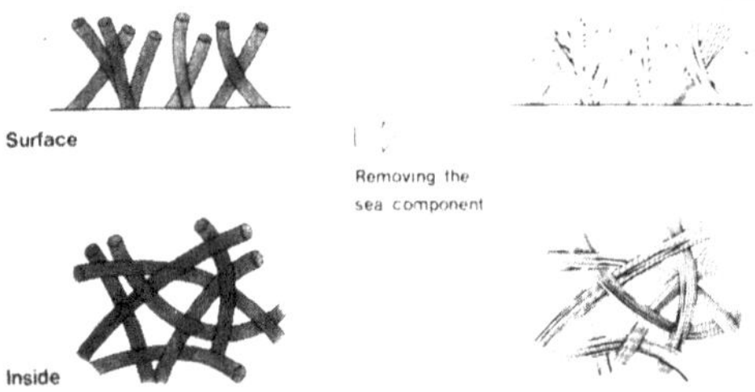

9.8 Processing and finishing of islands-in-a-sea fibers in non-woven sheet form.

9.9 Cross-sections of islands-in-a-sea fibers with various component ratios.

capable of producing a PET filament of 0.00009 denier (0.1 μm in diameter). Only 4.16 g of such a filament would be enough to stretch from the earth to the moon. The technology has been further extended to spin a multi-component conjugate fiber, and a suede-type artificial leather of high dyeability has been developed with a three-component conjugate extrusion where the two-component (polyester and nylon 6) ultra-fine fiber had a sheath-and-core structure. Many variations will be found in islands-in-a-sea spinning (Fig. 9.10), such as extremely-many-islands-component spinning for wiping cloths, non-circular cross-

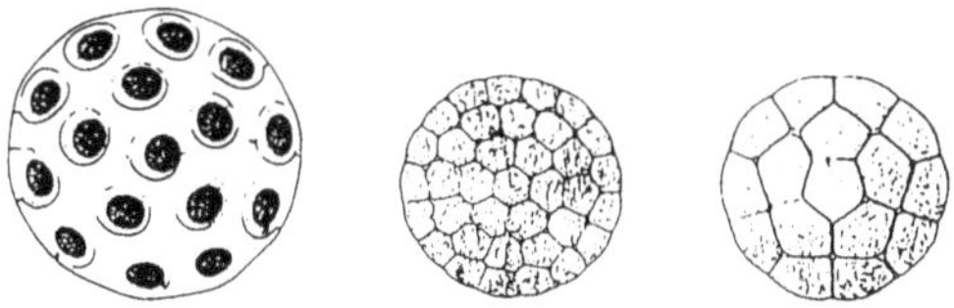

9.10 Islands-in-a-sea fibers with sheath core bicomponent islands (left) and with very little sea component (center and right).

sectional or non-even lateral surface island-component spinning, and blended island-component spinning.

9.3.2.2 Splitting type or separation type spinning[13,14,22,23,33]

This type of spinning aims to utilize the second component in the final product as well as the first by splitting the two components mechanically instead of removing the second component by dissolving. The fundamental principles of this technology will be found in the inventions by Breen[13] and Tanner[14] of Du Pont, although these inventors had no intention of applying their idea to the production of ultra-fine fibers. Since the commercial success of the artificial leather made from the alternately-arranged two-component fine fiber, this alternative spinning technology has been refined by Kanebo, Teijin and Toray to produce ultra-fine fibers.

Their ultra-fine spinning technology consists of the combination and separation of PET and nylon, the use of benzyl alcohol for effective shrinkage of the nylon component, and the establishment of dyeing technology for dyeing PET and nylon of different dyeing characteristics simultaneously with high colour fastness. These ultra-fine fibers have been developed mainly into moisture-permeable and water-repellent fabrics, along with suede-type artificial leather, silk-like fabrics and wiping cloths.

The ultra-fine fiberization is performed by a mechanical or chemical process in the splitting and separation types of spinning. Here the point is how many divisions of two components can be achieved. The commercial ultra-fine fibers are now produced by a specially designed spinneret. Toray and Kanebo employ a spinneret with a *-shaped cross-section, which is a modified version of that described in Breen's patent[13] (Fig. 9.11). The fiber, with a +-shaped cross-section, divides the second component into four wedges in this process. Kanebo has added another wedge component to each of the four wedges in order to increase the splitting probability in later processes. Its cross-section looks like a *-shaped conjugate fiber (Fig. 9.12 and 9.13).

Teijin produces a hollow-cylindrical conjugate fiber with a petal-shaped cross-section[24] as shown in Fig. 9.14. The basic design of the spinneret is similar to those of Du Pont[13] and Toray for the petal-shaped

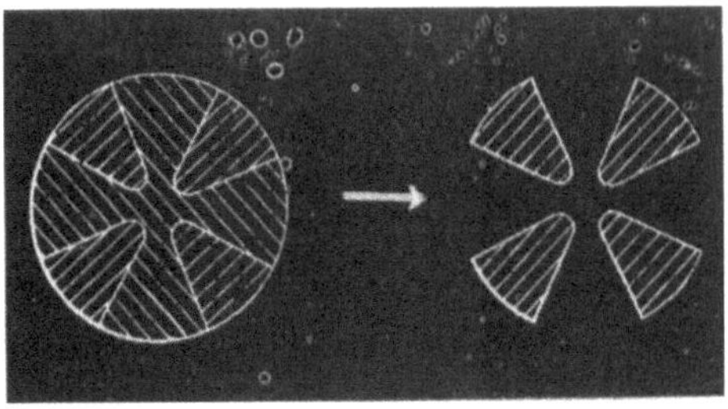

9.11 Petal-like conjugate fiber.

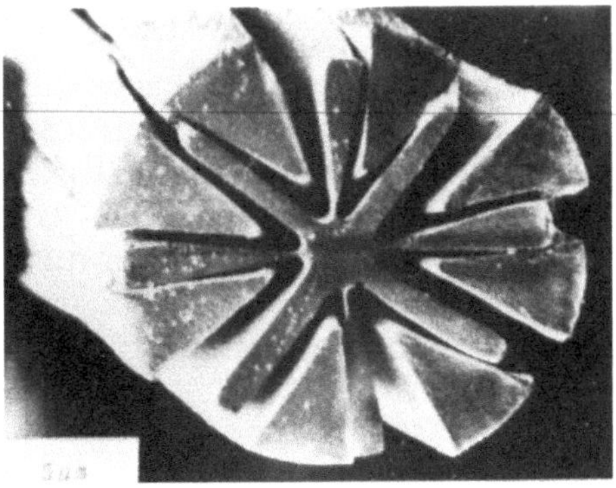

9.12 Petal-like conjugate fiber with 8 petals (Kanebo).

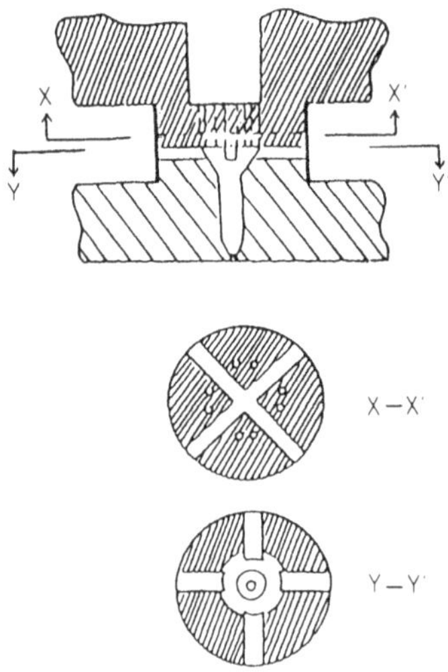

9.13 Spinneret hole arrangement for star-shaped conjugate fiber (Kanebo).

conjugate fiber. The hollow was introduced to avoid flattening and sharp edges in the resulting filaments even when the splitting number is increased. A good handle will be achieved when the spacing between ultra-fine filaments is large in the fabric. The hollow portion preserves the spacing between filaments after splitting, although the split filaments tend to arrange with the original cross-section before splitting because of the spatial restrictions in the fabric.

The spinneret is shown in Fig. 9.15. A proper choice of the spinneret and component polymers is vital in ultra-fine fiber spinning by this technique. PET and nylon 6 form the most popular combination found in the commercial products, since both components have a similar temperature range for extrusion and drawing. Fluffing during the drawing or weaving process can be avoided by improving the adhesion of two components, for example, by using the copolymer of PET and sodium sulfo-isophthalate as the PET component. Splitting during the spinning process can be reduced by increasing the spinning speed so that the PET and nylon exhibit similar shrinking behaviour.

9.3.2.3 Multi-layer type spinning

Liquids can be multi-layered by static mixers, which have been applied to multi-layer type spinning. Kanebo, Kuraray and Toray investigated this multi-layer type spinning technology. Kuraray introduced the first commercial multi-layer type ultra-fine fiber product (see Fig. 9.16 (a)). Here the two components, polyester and nylon 6, are spun into a conjugate fiber of multi-layered structure with an oval-shaped cross-section, which is micro-fiberized into filaments of 0.2–0.3 denier during the dyeing process.

Table 9.2 summarizes the characteristics of the above three methods (direct spinning, island-in-a-sea type and multi-layer, including separation

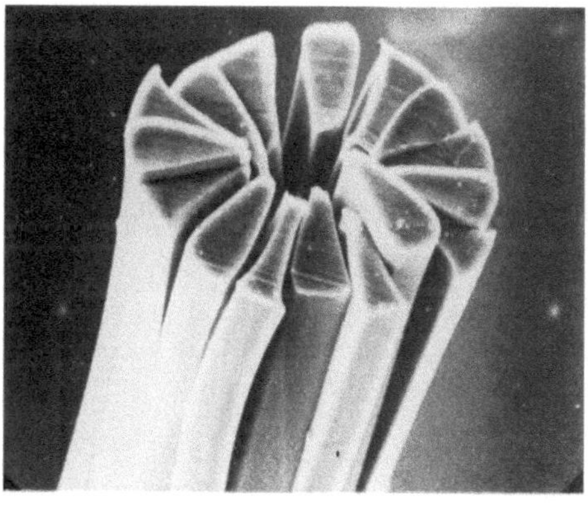

9.14 Petal-like hollow conjugate fiber (Teijin).

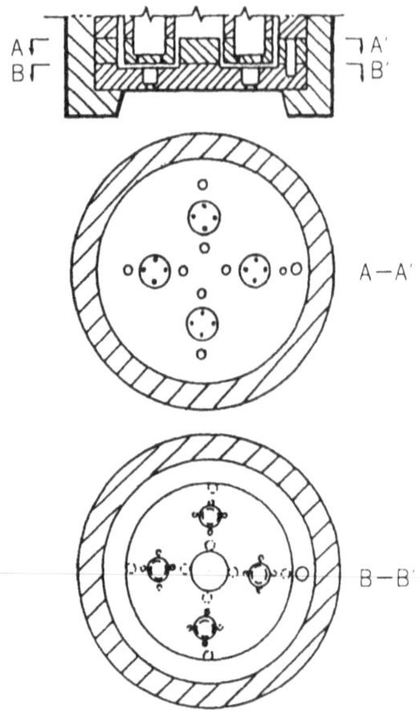

9.15 Spinneret for petal-like hollow conjugate fiber (Teijin).

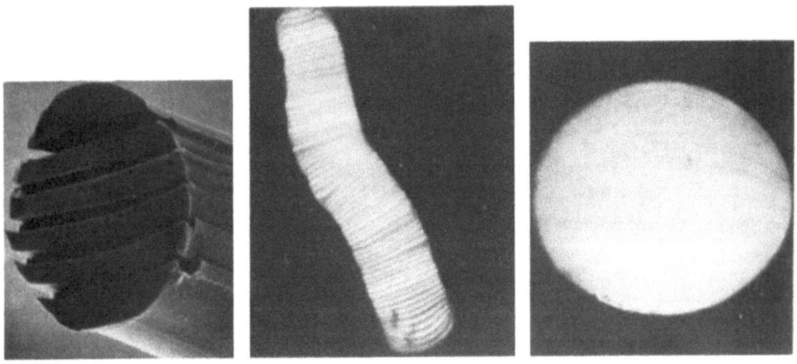

(a) (b) (c)

9.16 Oval multi-layer conjugate fiber (Kuraray and Toray).

or splitting, type) of ultra-fine continuous-filament fiber production and the typical properties of the products.

9.4 Random-type spinning

9.4.1 Melt-blowing or jet spinning[1,27]

This method is employed for the production of nonwoven fabrics of polypropylene ultra-fine fibers. The polymer melt is blown apart

Table 9.2 Comparison of characteristics of ultra-fine fiber of filament type according to production method

Filament	Direct spinning	Islands-in-a-sea type	Multi-layer type
Limit of the finest denier (after splitting)	>0.1 d	~0.0001 d Possible also to make non-circular cross-section	0.1d Flat cross-section
Method	Spinning and drawing	Remove a component by solvent	Physico-chemical splitting
Processing	Relatively difficult. Different from conventional fibers. Fluffy.	Relatively easy. Similar to conventional fibers. Less fluffy.	Relatively easy. Similar to conventional fibers. Fluffy.
Inter-filament distance in fabric	Small.	Can be controlled.	Small.
Hand.	Hard.	Soft.	Hard unless chemically treated to shrink.
Single-component fiber	Yes. Also possible to make multi-component fiber.	Yes (by dissolving). Also possible to make multi-component fiber.	No. (Only possible by also dissolving).
Dyeability/Color fastness	Easy dying. Poor color development.	Easy dying. Poor color development.	Poor color-fastness.

immediately after extrusion by an air-jet stream in this method, so it is sometimes termed 'jet spinning'. Thus, this method is an application of spraying technology rather than true spinning. The melt viscosity is lower than that of the conventional polypropylene used for melt spinning, and an appropriate range of melt viscosity should be chosen to achieve the required result. The spinneret has a sharp edge and the jet stream blows off the extruded polymer melt into ultra-fine fibers that are collected as nonwovens. New technology has been developed for the production of electrets on these polypropylene nonwovens to meet the requirements for a clean environment. Chapter 4 gives further details.

9.4.2 Flash spinning[1]

Flash-spun fiber (shown in Fig. 9.17[33]) is counted as an ultra-fine fiber. Figure 9.17 is a photograph of a fiber network obtained by spreading a single stream of fiber spun from one spinneret hole. The filament thickness varies from 0.01 denier to 10 denier, and the average denier per filament is approximately 0.1–0.15 denier. The filament cross-section is non-circular, and some filaments contain micro-bubbles. This technology was found by Du Pont accidentally while examining the explosion behavior of organic solvents for safety research. As an example of the process, polyethylene is dissolved in hydrocarbon or methylene chloride, heated under pressure, and jetted from a nozzle into a micro-networked fiber termed a 'plexifilament'. Ethylene chloride and fluorocarbon have been used as solvents, but they are now being replaced with other solvents which will not destroy the ozone layer. (It is said that Du Pont has already developed a new solvent.) This technology was first aimed at producing pulp for synthetic paper, but has been converted to produce wrapping materials including wrappings for domestic use and envelopes. The spinning speed is too high to reel in the product in the form of fiber, so the product is made into a sheet.

The spinneret for flash spinning and the state of the spinning solutions at each stage are shown in Fig. 9.18 and 9.19. Polymer is dissolved in liquid gas and made into transparent (homogeneous) solutions at high temperature under high pressure. The spinning solution is then liquid–liquid phase-separated to make a turbid solution, which is jetted out of a nozzle into air to form a fibrous network. The homogeneous polymer solution (prepared by extrusion or in autoclave, and kept at a high temperature under high pressure) is extruded through point *a* and the orifice in the pressure-reducing chamber *b* (see Fig. 9.18). Since the pressure at point *a* is reduced in chamber b, the polymer solution separates into two phases, and is then jetted into air through the nozzle. In the process from *b* to c, the solvent (liquefied gas) is gasified instantly and produces a jet stream. The temperature drops to room temperature, and the dissolved polymer is solidified and drawn to form a fiber of high strength. Since the micro-phases of solvent are dispersed in the polymer gel which expands explosively at the time of jetting, the resulting fiber is composed of an extremely thin network of non-circular cross-section fibrils like a cobweb.[33]

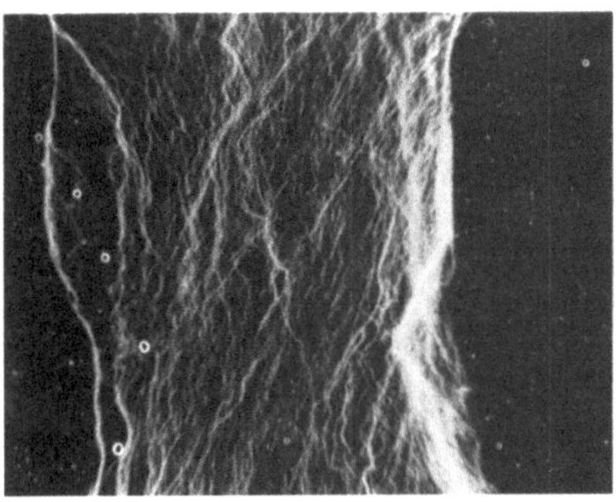

9.17 Flash-spun fiber (about 100 filaments; Asahi Kasei).

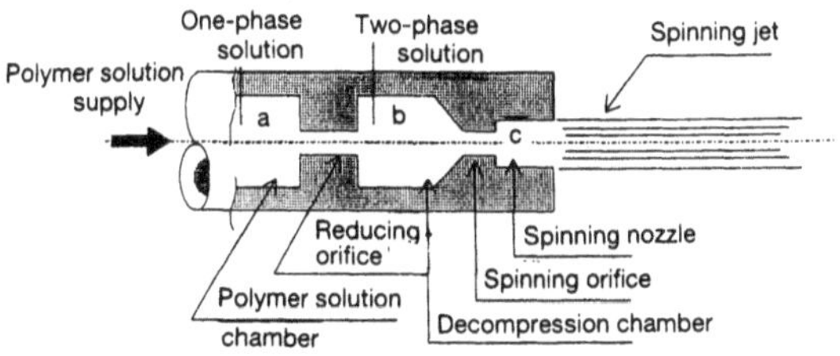

9.18 Conceptual diagram for flash-spinning nozzle.

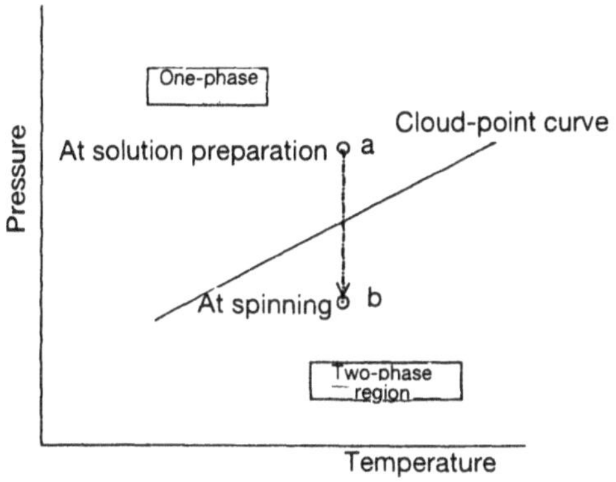

9.19 Phase diagram illustrating flash-spinning principle.

9.4.3 Polymer blend spinning[1,32]

In this method the conjugate fiber is produced by extruding and drawing a blended polymer melt of two components. The arrangement of the dispersed and non-dispersed (matrix) components is determined by the mixing ratio of the components and their melt viscosities. UCC and others found that discontinuous ultra-fine fibers could be obtained by removing the matrix component while investigating gut extrusion. Fukushima *et al.*[32] of Kuraray successfully applied this method in the production of artificial leather. A conventional spinning facility can be easily converted for this type of polymer blend spinning by adding a mixer–extruder. Here, the fiber fineness cannot be controlled and the fiber often breaks during spinning, although the spinning stability is strongly dependent on the combination of polymers. Since the dispersed polymer phase is drawn to yield ultra-fine fibers, no continuous-filament type of ultra-fine fiber is produced at present by polymer blend spinning.

9.4.4 Other random-type spinning[1–8]

Other random-type spinning processes for ultra-fine fiber are:

1 Ultra-centrifugal spinning, where ultra-fine fiber is spun like a cotton candy.
2 Fibrillation by beating, where a fiber or film is beaten to fibrillate it.
3 Fibrillation by turbulent flow, where a polymer in solution is coagulated in a zone of turbulent flow.
4 Burst spinning (blown sheet method), where a blowing agent or gas is introduced into the polymer to burst it apart and form ultra-fine staple fibers.
5 Surface dissolving method, where the surface of PET fibers etc. is dissolved in alkaline solution and the fibers are thereby made thinner.
6 Ultra-fine fiber formation is also effected by whisker production, by reducing weight by carbonization, by tack-spinning, and by high-pressure water flow techniques.

9.5 Conclusion

Ultra-fine fiber is not a speciality fiber but can be considered to be one of the fundamental fiber materials today. It is expected to open up new fields of fiber application as demonstrated in Fig. 9.20. Ultra-fine fiber spinning technology will be developed further towards its limits, and will be applied in other fields of technology.

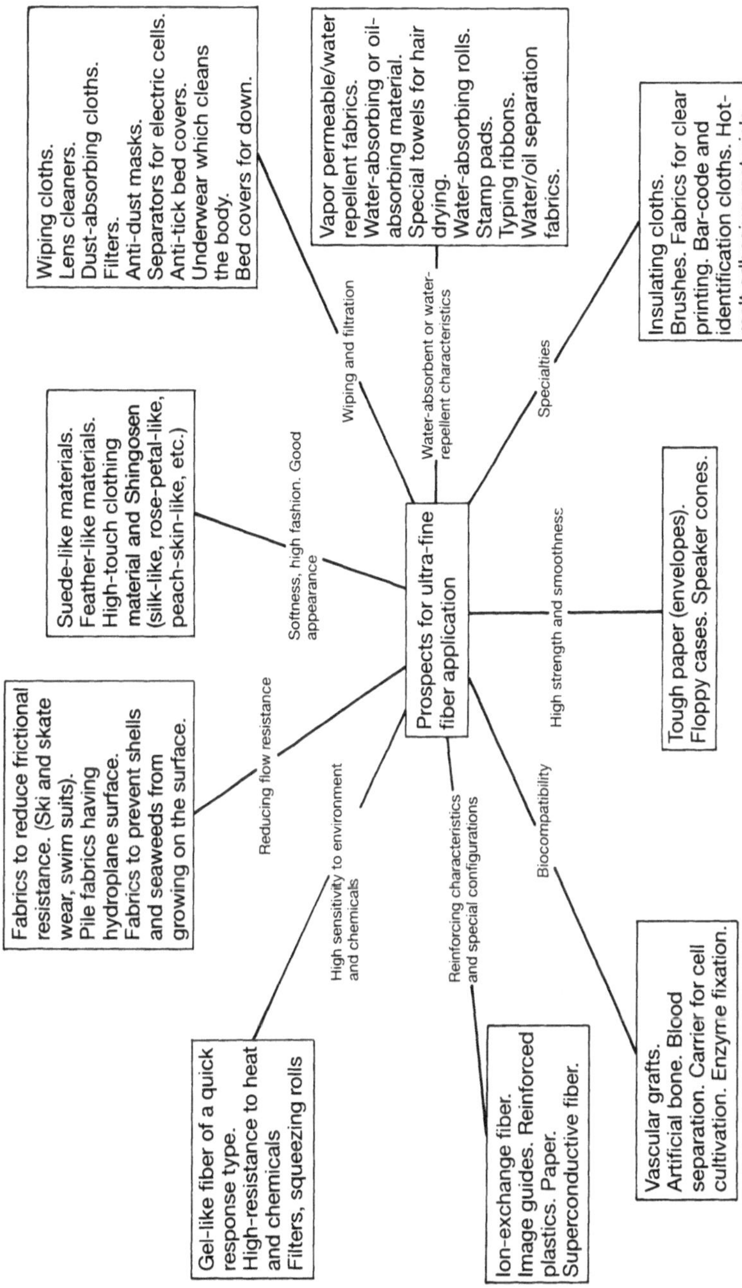

9.20 Prospective applications for ultra-fine fibers.

References

1 Okamoto M, On ultra fine fibers, *Sen-i Gakkaishi*, **32**, P-318, 1976.
2 Okamoto M, Ultra-fine fiber and its application, *JTN*, p.94, Nov. 1977; JTN, p.77, Jan. 1978.
3 Okamoto M, Progress of artificial leather, *Gendai Kaguku*, p.52, Aug. 1981.
4 Okamoto M, Reprints from *Japan-China Bilateral Symposium on Polymer Science and Technology*, p.256, 1981.
5 Okamoto M, Ultra-fine fibers, *Kagaku-to-Kogyo*, **36**, 521, 1983.
6 Okamoto M, *Chemiefasern Textilindustrie*, **29/81**, 30 and E4, 1979.
7 Okamoto M, *Chemiefasern Textilindustrie*, **29/81**, 175 and E25, 1979.
8 Okamoto M, American Association for Textile Technology (AATT), Charlotte, NC, U.S.A., Oct. (1988)
9 Heidenreich I, Microfibers – a passing fashion or the standard of the future?, *30th International Conference on Chemical Fibres, Dornbirn, Austria* 1991.
10 Davies S, *Textile Horizons*, **8**, 49, Apr. 1988.
11 Isaacs III M, *Textile World*, **140**, 48, Mar. 1991.
12 Maeda Y, *Sen-i Kikai Gakkaishi*, **42**, 446, 1989.
13 Du Pont, Jap. Pat. (Examined) 1964-29,636.
14 Du Pont, Jap. Pat. (Examined) 1964-933.
15 Kanebo, Jap. Pat. (Examined) 1972-49,766.
16 Dupeuble J C, America's Textiles International, p.48, May 1991.
17 Davidson W A B, America's Textiles International, **20**, p.44, Apr. 1991.
18 Koslowski H-J, *Chemiefasern Textilindustrie*, **40/92**, 73, 1990.
19 Fujimoto K, Iohara K, Owaki S and Murase Y, *Sen-i Gakkaishi*, **44**, 477, 1988.
20 Toray, Jap. Pat. (Examined) 1968-14,184; 1968-19,604; 1968-7,411.
21 Toray, Jap. Pat. (Examined) 1969-18,369; 1973-22,126; 1984-14,570; 1972-26,723; 1987-25,763; 1973-22,127; 1971-3,816; 1972-9,850; 1972-42,847; 1975-13,849; 1969-13,208; 1969-3,505; 1969-2,166.
22 Toray, Jap. Pat. (Examined) 1983-12,367; 1988-3,963; 1987-25,766; 1983-31,402; 1986-60,164; 1983-4,087; 1982-49,651; 1982-34,368.
23 Matsui M, *Sen-i Kikai Gakkaishi*, **34**, 319, 1981; *Sen-i Kikai Gakkaishi*, **41**, 122, 1988; *Kako Gijutsu*, **24**, 14, 1989; also in *High Performance Polymer Alloys*, ed. Society of Polymer Science and Technology, Tokyo, 1991.
24 Teijin, Jap. Pat. (Examined) 1978-10,169.
25 Imai S, *Sen-i Gakkai Preprints*, 278, 1990.
26 Mizuki T, *Polymer Preprints, Japan*, **40**, 1453, 1991.
27 Kuraray, Jap. Pat. (Examined) 1990-289,107; Toray, Jap. Pat. (Examined) 1990-289,162; Asahi Kasei, Jap. Pat. (Laid open) 1980-90,663; 1976-6,741; Esso Research & Engineering, Jap. Pat. (Laid open) 1974-48,921.
28 Asahi Kasei, Jap. Pat. (Examined) 1987-35,481; 1987-50,566; 1988-526; 1988-8,206; 1988-17,921.
29 Teijin, Jap. Pat. (Examined) 1983-37,408; 1987-11,085.
30 Unitika, Jap. Pat. (Laid open) 1980-128,007; 1981-101,907; 1981-107,006; 1982-11,208; Jap. Pat. (Examined) 1983-44,763; 1985-25,447; 1991-12,419; 1991-12,420.
31 Mitsubishi Rayon, Jap. Pat. (Laid open) 1976-119,826; 1978-122,815; 1979-147,203; 1982-167,410.
32 Fukushima, O and Kogame K, *Sen-i Gakkaishi*, **39**, P-452, 1983; also in

Fundamentals and Applications of Polymer Alloys, ed. Society of Polymer Science and Technology, Kagaku-Dojin, Tokyo, 1981.
33 Asahi Kasei, Teijin, Unitika, Kuraray, Mitsubishi Rayon, Toyobo, *private communications*, June 1991.
34 Nakayama Y, *Sen-i Gakkaishi*, **47**, P-347, 1991.

10

Spinning of optical fibers

Katsuhiko Shimada
Mitsubishi Rayon Co. Ltd, Toyama, Japan

10.1 Introduction

Since signal transmission by light has been regarded as important for the progress of the communication media, attention has been paid to optical communication using light emitting and receiving devices, and particularly to optical fiber as the central component.

An optical fiber is composed of a light-transmitting, transparent core material coated with a material with lower refractive index. They are classified into three groups according to the types of core material, quartz, multi-component and plastic optical fibers (POF). The quartz optical fiber is used for long-distance optical communication including public trunk lines because its transmission loss is as low as 1 dB/km or under. The multi-component optical fiber is used for middle-distance communication of 1–2 km including local area networks (LAN) in plants and fiberscopes. The glass optical fiber has a shortcoming in that its processing requires great skill because it is expensive, brittle and hard to process. Meanwhile, the plastic optical fiber has advantages in that its inexpensive, flexible, light and easy to process, though the transmission loss is as high as 120–130 dB/km. It is, therefore, widely used for short-distance wave guides including optical sensors, optical transmission in equipment, mobile optical communication and display.

Although a plastic optical fiber using polystyrene as the core material was commercialised early in the 1960s, the transmission loss was as high as 5000 dB/km. DuPont launched a plastic optical fiber using polymethylmethacrylate (PMMA) as the core material late in the 1960s. Du Pont were followed by Mitsubishi Rayon who produced their own PMMA plastic optical fiber in 1975. It was, however, only used in automotive light guides, signboards and display systems because the transmission loss was still as high as 1000 dB/km. From active studies and theoretical analyses for improving the transmission loss, Nippon Telegraph and Telephone Public Corporation (now NTT) indicated in 1985 that the theoretical limiting transmission loss of the plastic optical fiber using PMMA as the core material, was 34.9 dB/km at a wavelength of 568 nm.[1] Mitsubishi Rayon had conducted research into improving the transmission loss on the industrial scale and decreased it to 300, 150 and 120 dB/km at a wavelength of 650 nm in 1977, 1982 and 1989, respectively. Since long-distance optical communication (150–180 m) was realized by this, optical signal transmission has rapidly spread using plastic optical fibers in a wide range of applications.

This chapter outlines the process for manufacturing plastic optical fibers, the technique for improving the transmission loss and the latest product with a multiple optical element.

10.2 Structure of plastic optical fiber

The structure of optical fiber is classified by the transmission methods into the step index (SI) and graded index (GI) types. These transmission types have the specific features shown in Fig. 10.1. Commercial plastic optical fiber is of the SI type. The GI type has not been commercialized yet. It is interesting that the transmission loss and transmission bandwidth of the GI type plastic optical fiber are 134 dB/km and 360 MHz.km, respectively.[2] Recently, a single mode plastic optical fiber has been proposed and manufactured on an experimental basis[3] and the types are being rapidly varied. The structural dimensions of the plastic optical fiber using PMMA as the core material are being standardized in such a way that the core diameter and sheath thickness of a plastic optical fiber 1000 μm in outer diameter are 980 and 10 μm, respectively. The area ratio

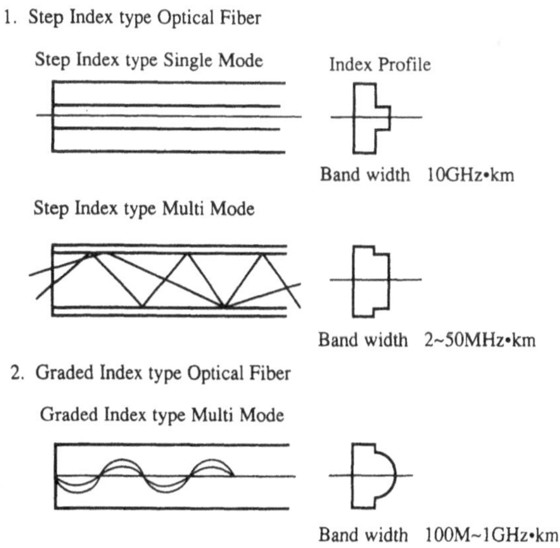

10.1 Distribution of refractive index in optical fibers and their transmission bandwidths.

of the core is so large that it is characterized easily by connecting an optical fiber to a light source and another optical fiber in alignment.

10.3 Material of plastic optical fibre

The materials of both the core and sheath of the plastic optical fiber require high transparency. Although commercially available PMMA and polycarbonate (PC) are generally used as the core material, silicone and thermosetting resins, MMA-methacrylic anhydride copolymer, MMA-maleic anhydride copolymer, MMA-methacrylicimide copolymer and ionomers are being investigated. Since the refractive index of the sheath material should be lower than that of the core material, fluoroplastics including poly (vinylidene fluoride), Teflon FEP, Teflon AF, fluorinated methacrylate and fluroinated polycarbonate are used as the sheath material. Polymethylpentene is used as the sheath material for a plastic optical fiber with a relatively high refractive index of 1.59 using fluorinated polycarbonate as the core material.

10.4 Causes of transmission loss of plastic optical fiber

The causes of the transmission loss of the plastic optical fiber are listed in Fig. 10.2. They are classified into the loss inherent in the material itself and the external loss from the manufacturing technology. Kaino[1] investigated the internal losses theoretically including the losses on absorption and scattering for a plastic optical fiber using PMMA as the core material and estimated that the limit of the internal loss was 106.2 dB/km at a wavelength of 650 nm in the visible LED.

The typical transmission losses as a function of wavelength of Eska Extra (produced by Mitsubishi Rayon) which has the lowest transmission loss among the current commercially available plastic optical fibers, are illustrated in Fig 10.3. Although the transmission loss at a wavelength of 568 nm is 60–70 dB/km which is 30–40 dB/km smaller than the limiting loss, it is considerably lower than the value for the product manufactured

- Intrinsic
  - Absorption
    - Higher harmonics of C–H vibration
    - Electronic transition
  - Scattering
    - Rayleigh scattering
- Extrinsic
  - Absorption
    - Transition metals
    - Organic contaminants
  - Scattering
    - Dust and microvoids
    - Fluctuation in core diameter
    - Orientational birefringence
    - Core-cladding boundary imperfections

10.2 Transmission loss factors for POF.

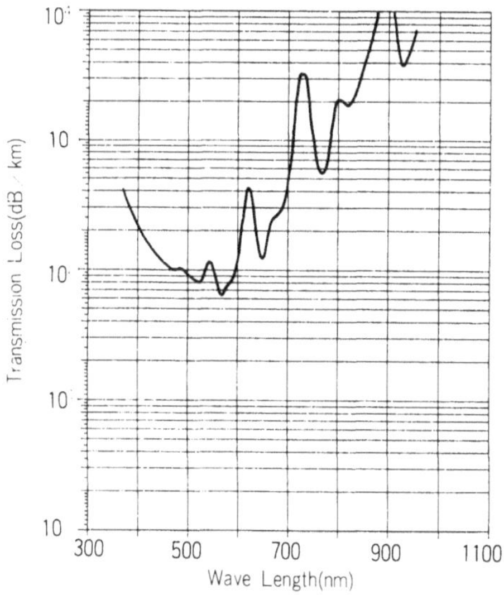

10.3 Transmission of Eska Extra.

on the industrial scale. This transmission loss is 120 dB/km at a wavelength in the vicinity of 650 nm in the visible LED used for communication which is close to 106.2 dB/km, the limiting loss obtained by Kaino. The process for commercially manufacturing a low-transmission loss plastic optical fiber using PMMA as the core material will now be described.

10.5 Process for manufacturing plastic optical fiber using PMMA as core material

The schematic flow diagram of the process for manufacturing plastic optical fiber using PMMA as the core material is illustrated in Fig. 10.4. Although suspension and bulk polymerizations are applicable to the process, the latter lowers the transmission loss because the usages of the polymerization initiator and the molecular weight modifier are substantially less. The transmission loss is, however, increased by pelletising the core material made by the bulk polymerization and by melt spinning to produce the plastic optical fiber. Mitsubishi Rayon started to produce high-purity PMMA for the first time by an industrial continuous bulk polymerization of MMA in 1970. Based on the actual results obtained by the operation of the process, Mitsubishi Rayon put Eska Extra, the plastic optical fiber for communication, on the market in 1983 by combining the continuous polymerization technology of MMA with the melt spinning technology of polyester and polypropylene fibers. The transmission loss of this product was half that of conventional plastic

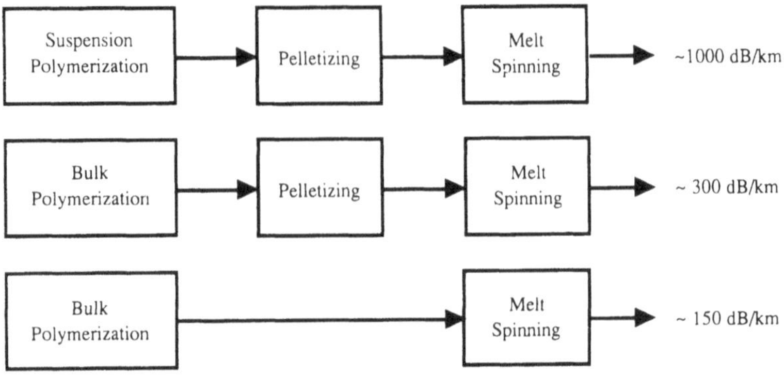

10.4 Processing methods and resulting transmission losses of POF with PMMA as a core

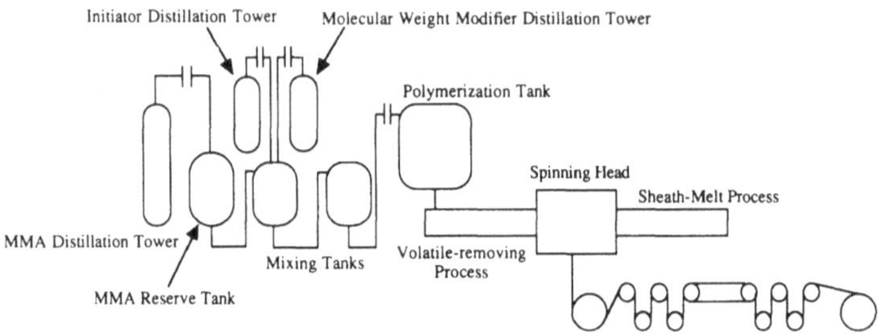

10.5 Flow chart of POF production.

optical fibers and the transmission distance was increased to as long as 150 m. The process flow diagram of the plastic optical fiber is illustrated in Fig. 10.5.

MMA contains a polymerization inhibitor, which should be removed by distillation. When a hydroquinone derivative, generally used as the polymerization inhibitor, remains in the final product, it absorbs visible wavelengths.

The dissolved oxygen in the monomer reacts with the monomer or molecular weight modifier in the polymerization process and forms a peroxide which easily colours PMMA. MMA peroxides contained in the monomer should also be removed. These peroxides can be removed by pretreatment with a reducing agent followed by distillation. Various methods for removing compounds contained in the monomer affecting the visible wavelengths have been described.[4]

After removing the polymerization inhibitor and molecular weight modifier from the monomer by distillation and also thoroughly removing impurities by vapor and liquid-phase filters, the monomer is ready for polymerization. A vapor phase filter using hollow yarns with an aperture of 700Å is effective.[5] Various other purification processes for the monomer have also been proposed.[6]

As mentioned above, the bulk polymerization process is most suitable for the manufacture of plastic optical fibers because polymerization additives, other than the polymerization inhibitor and molecular weight modifier, are not used.

Mitsubishi Rayon has developed a process for manufacturing PMMA involving continuous bulk polymerization of MMA followed by a process for removing volatile matter such as unreacted monomer. PMMA is manufactured in this newly-developed process in such a way that MMA is polymerized using tens of ppm of the polymerization initiator and molecular weight modifier at 130–180°C for several hours at a yield of approximately 60 wt% and the volatile matter, including unreacted monomer and molecular weight modifier, is then removed. This polymerization process is an integral part of the manufacturing process of Eska Extra.

It is important for stable polymerization to produce a uniformly mixed blend of monomer and polymer. Since the termination reaction is retarded by the increase of the viscosity in the polymerization reaction, the polymerization sometimes proceeds rapidly by the gel effect into a runaway reaction. It is, therefore, important to remove the heat of polymerization effectively by reducing the viscosity to below a specified value and uniformly stirring the polymerization product. Since variations in the degree of polymerization and the viscosity of the blend cause variations in the blend transportation and volatile-removing process, the transmission loss of the plastic optical fiber can vary.

The polymerization temperature is preferably kept low so that there is no gel effect. Oligomers including MMA dimer are produced as by-products by increasing the polymerization temperature. Since the oligomers are so thermally unstable that they are easily coloured, the absorption of the plastic optical fiber is increased.[7]

Important features of the volatile-removing process are to remove the volatile matter including monomer, at a temperature as low as possible, to avoid locally heating the polymer by high shearing, to prevent any dead space from forming in the equipment and to inhibit the generation of impurities from the equipment. Two processes are proposed for the volatile-removing process as illustrated in Fig. 10.6. The special features of these two processes will now be described.[8,9]

10.5.1 Extruder with degassing hole

In a conventional extruder with a degassing vent, an upward degassing vent is connected to the evacuation line ahead of the blend-feeding hole and directly connected to the extruder cylinder. In the newly-developed extruder, however, the degassing vent is fixed behind the blend-feeding

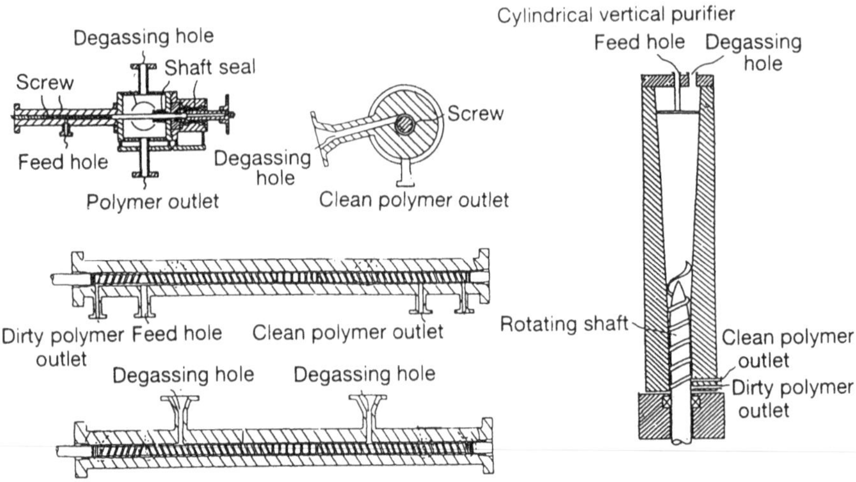

10.6 Schematic view of volatile-removing process.

hole so that part of the blend is sent to the rear of the shaft of the screw through an inverse pitch screw channel, so that the volatile matter is steadily discharged together with the blend from the system without mixing the impurites generated by rubbing at the shaft seal with the polymerization product. Since the direction of the degassing hole is horizontal, the polymer discharged from it together with the monomer does not return to the screw channel. Impurity-free polymer is, therefore, continuously extruded from the end of the extruder.

10.5.2 Cylindrical vertical purifier

The blend is fed from a small hole at the end of the feeding pipe against the upper inside wall of the cylinder of the volatile-removing equipment. The temperature of the inside wall is controlled in advance at a temperature such that the viscosity of the polymer is kept at 5000 poise or less. The syrup fed against the heated wall flows down, spreading on the wall in the form of film. Since the cylinder is evacuated from the degassing hole, volatile matter contained in the blend is effectively removed. The polymer free from the volatile matter is fed to a polymer tank located in the deaeration zone while maintaining the free level at a constant height in the lower part of the cylinder. The volatile matter is removed by gently stirring the syrup to release trapped bubbles under the free level. The polymer is then transported downward under pressure into a screw and discharged from its outlet. The polymer is prevented from contamination with the discharge from the under shaft seal by discharging it from a separate outlet. Since this purifier is provided with a vertical rotating shaft, there is little vibration and the polymer is never contaminated by abrasion.

Particularly important in the volatile-removing process is the dissolution of metals from the metallic pipe, screw and cylinder. Transition metals removed from the equipment by corrosion cause transmission losses in the plastic optical fiber. The effect of the transition metals on the transmission loss is shown in Table 10.1. The effects of Co and Cr are particularly large in the visible region.[10]

10.6 Spinning technique for plastic optical fiber using PMMA as core material

The core material melted in the volatile-removing process is supplied to the core-sheath composite spinning nozzle through a constant-volume gear pump. A schematic drawing of the spinning nozzle is illustrated in Fig. 10.7. Attention has to be paid to the following matters in the spinning of plastic optical fiber:

1 Avoiding long residues of the polymer in piping and equipment
2 Preventing extraneous matter being generated from the equipment
3 Designing a high-level spinning nozzle.
4 Optimizing the viscosities of molten core and sheath materials.

Table 10.1 Influence of transition metals on transmission loss: amounts required to give a 10dB/km loss

Wavelength	Co	Cr	Mo	Fe	Cu	Ni
630nm	2	10	200	500	30	70
850nm	*	*	*	120	30	30

Units ppb, except * > ppm

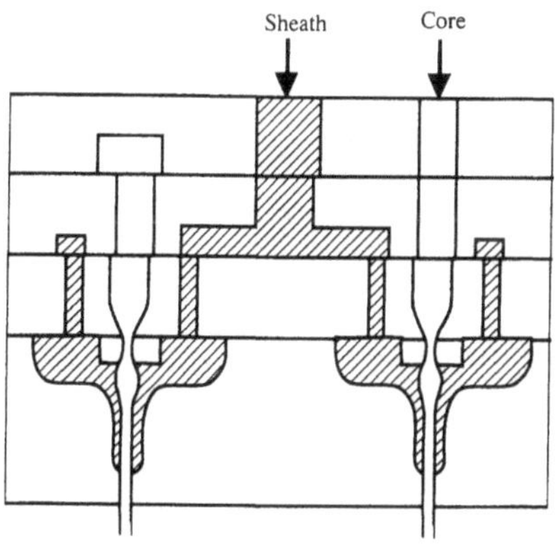

10.7 Spinning nozzle for conjugate spinning.

It is necessary to shorten the length of the piping after the polymerization process as much as possible and to electropolish the inside surface of the pipe to reduce the surface resistance. It is also necessary to increase the bending radius of the pipe as much as possible. It is important to conduct dead-end polymerization to prevent the polymer from staying in the piping end of the equipment. As illustrated in Fig. 10.8, dead-end polymerization is a process in which polymerization does not proceed after a specified time has passed. Even if any dead reactant space remains present in dead-end polymerization, the reactant is not polymerised there in practice.

For preventing the generation of extraneous matter from the equipment, the volatile-removing equipment has a mechanism for discharging contaminant polymer from the shearing part and degassing hole as described in section 10.5. It also has a constant-volume gear pump which steadily discharges the contaminant produced from the rotating-shearing part.[11]

The transmission loss caused by imperfect fibrous structure, including irregular interface and unequal core diameters, is mainly derived from the spinning process. Optimum design of the spinning nozzle and adaption of the fluidity of polymer to the spinning conditions are required for reducing the loss.

The important features of the processing at the spinning nozzle are shown in Table 10.2. Since the transmission loss of the plastic optical fiber is increased by the color developed by heating the core material, it should be spun at as low a temperature as possible. The melt viscosity of the core material is, however, raised to tens of thousands of poise at such a low temperature. The operational conditions including molecular weight, temperature, inner diameter of nozzle and output should be optimized by keeping the shear stress in the nozzle at 10^6 dyne/cm^2 or less to prevent the development of melt fracture.

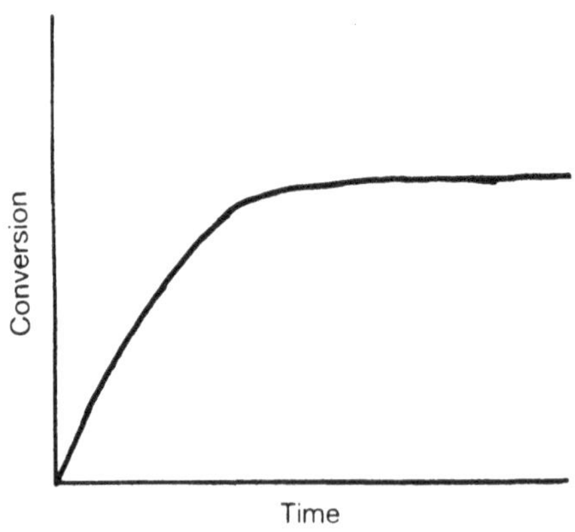

10.8 Dead-end polymerization: conversion with time.

Table 10.2 Methods of improving the processing nozzle

Processing region	Fiber shape	Requirement
Inner wall of nozzle	Imperfections of core-sheath inner surface Fluctuation in outer diameter	Improvement of wall smoothness
Core-sheath junction	Core, sheath concentricity Imperfections of core-sheath surface	Improvement of accuracy of nozzle Improvement of accuracy of nozzle plate setting
Nozzle length	Imperfections of core-sheath inner surface Fluctuation in outer diameter	Longer Shorter

Although composite melt spinning is effectively used for a composite fiber and multilayer film, the instability of the shape of the interface is interesting. A low viscosity material generally encapsulates a high viscosity material during passage through a circular die because the low viscosity material migrates to the high shear region at the die surface.[12,13] Since the smoothness of the interface between the core and the sheath is important, the melt viscosities of core and sheath should be optimized. Sectional photographs of plastic optical fibers taken whilst keeping the melt viscosity ratio of core to sheath at 1 and 2 are shown in Fig. 10.9. The interface between the materials at the melt viscosity ratio of 1 is uneven, while that at the melt viscosity ratio of 2 is smooth. Figure 10.10 illustrates the relationship between the melt viscosity ratio of core to sheath and the transmission loss of the plastic optical fiber. Although the transmission loss is decreased by increasing the melt viscosity ratio up

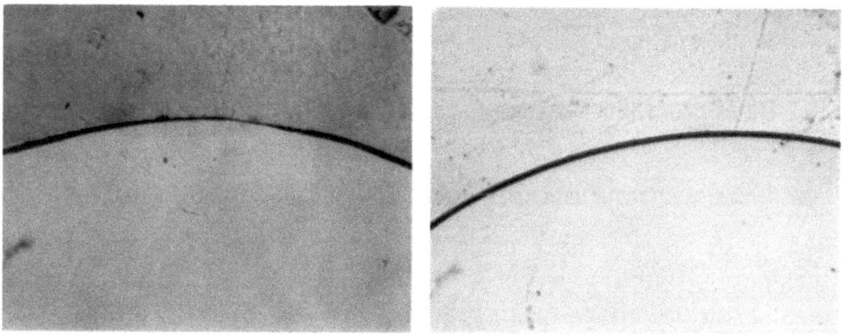

(a) (b)

10.9 Relationship between core and sheath, melt viscosities and cross-sections of POF. Core: PMMA. Sheath: Florinated methacrylate polymer. Melt index: 230°C, 3.8kg. Core–sheath viscosity ratio: (a) 1; (b) 2.

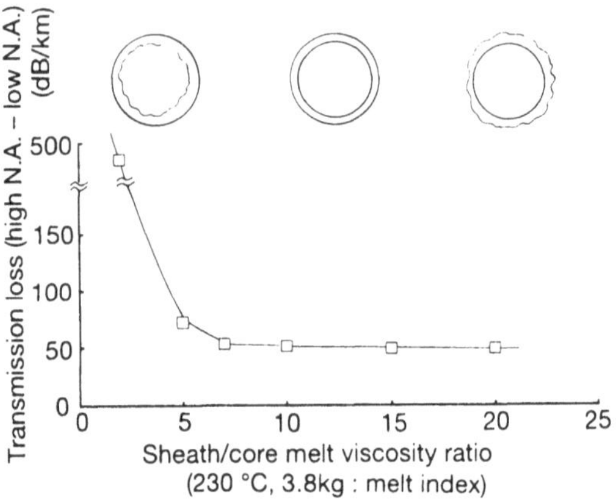

10.10 Relationship between sheath/core melt viscosity ratio and transmission loss.

to 10 or higher, the variability of the outer diameter of the plastic optical fiber is increased by decreasing the melt viscosity of the sheath too much. The conceptual relationship between the melt viscosities or core and sheath and the sectional shape of plastic optical fiber is shown in Fig. 10.10.

Research into reducing the transmission loss of plastic optical fiber has been energetically pursued and values for commercially available plastic optical fiber products are as low as 120 dB/km at a wavelength of 650 nm. Since the value for the plastic optical fiber using PMMA as the core material should still be capable of reduction by a further 20 dB/km, improvements in the process and material are expected. For further decreases in the transmission loss of the plastic optical fiber, the development of transparent materials including floropolymers is expected. When these are developed, the transmission loss may theoretically be reduced to 5 dB/km.[1]

10.7 Other processes for manufacturing plastic optical fiber

10.7.1 Process for manufacturing plastic optical fiber using PC as core material

The plastic optical fiber using PC as the core material has a heat resistance of 120°C or higher which is 30°C or so higher than that of the plastic optical fiber using PMMA as the core material. After commercialization of this plastic optical fiber by Mitsubishi Rayon in 1986, Fujitsu, Teijin Kasei, Idemitsu Petrochemical and Asahi Kasei have also successfully commercialized it.

PC polymer is dissolved in an organic solvent containing methylene chloride, and unreacted substances and by-products are removed from the solvent solution by washing. The polymer is recovered by removing the solvent using a spray drying method. This recovered polymer is usually pelletized by melt extrusion at a temperature higher than the crystalline melting point of 245°C, but the polymer thermally degrades at 280–320°C. Even though pellets of such a PC polymer containing thermal decomposition products can be used for the melt spinning of plastic optical fiber, high-transmissibility product cannot be obtained.[14] It is, therefore, necessary to directly carry out the melt spinning of polymer recovered from the polymerization process without pelletizing it to produce a plastic optical fiber with high transmissibility. It is also necessary to inhibit crystallization of polymer during the melt spinning. Since the crystallization point of PC reaches its maximum at approximately 190°C, the molding temperature should be 210°C or higher.

10.7.2 Process for manufacturing plastic optical fiber using organic silicone as core material

The plastic optical fiber using organic silicone as core material is characterized by flexibility and resistance to heat and chemicals. Sumitomo Denko has commercialized such a fiber composed of silicone rubber as the core and Teflon FEP as the sheath. Three manufacturing processes are as follows:

1 A mixture of vinyl alkyl siloxane and a platinum catalyst is filtered and the filtrate is injected into a hollow tube of a tetrafluoroethylene/hexafluoropropylene copolymer (FEP) under vacuum followed by thermal polymerization.[15]
2 Using a mixture of liquid siloxane polymer and a hydrogen chloroplatinate as the core material, and liquid siloxane polymer with lower refractive index that that of the core materials as the sheath material, both materials are fed at the same time through nozzles 2 mm and 4 mm in diameter, respectively (Fig. 10.11) to a thermal cross-linking stage in a heater.[16]
3 A fiber of the core material is manufactured in the same way as in 2 and the sheath material is coated by dip coating.[17]

10.7.3 Process for manufacturing plastic optical fiber using thermosetting resin as core material

Since the plastic optical fiber using thermosetting resin as core material has good retention of shape at high temperature, Hitachi Densen has placed it on the market. It is composed of a thermosetting resin as the core material and Teflon FEP as the sheath material. The monomer is fed through a pump into a stainless steel tube lined with Teflon. It is

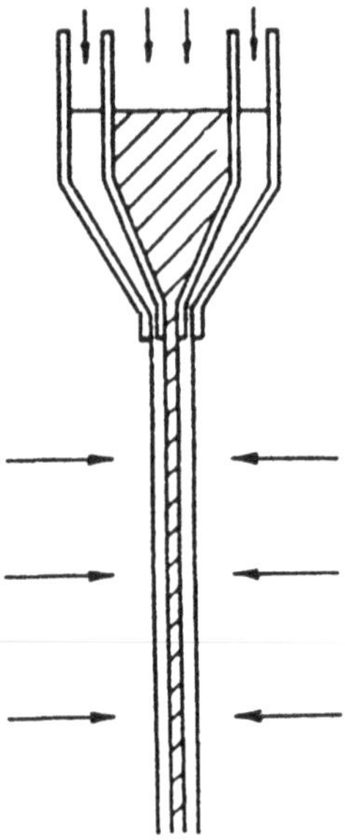

10.11 Process for manufacturing organic-silicone core POF.

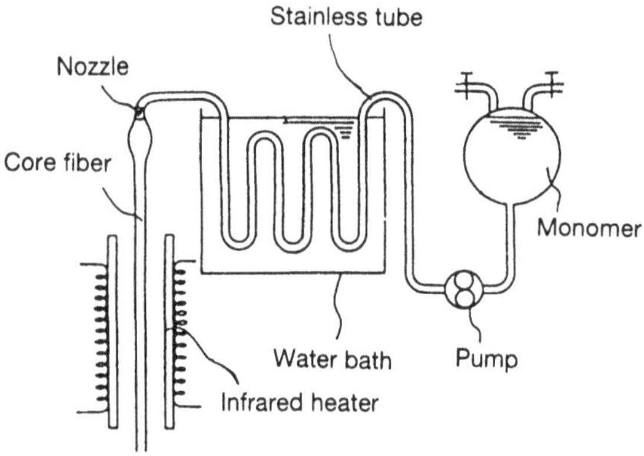

10.12 Process for manufacturing thermosetting resin core POF.

polymerized in a hot water tank so that the viscosity of the polymer reaches 10^4 poise. A fibrous resin is extruded from the nozzle and heated by an infra-red heater to form the core (Fig. 10.12). The resin is then coated with a cladding material.[18]

10.8 Multi-picture element plastic optical fiber

Diverse applications of plastic optical fibers including light transmitters, optical sensors, optical branching fibers and image guides are expected to be developed. Sheet and block-type multi-picture element plastic optical fibers have been developed. Since these plastic optical fibers are presently manufactured by accurately arranging a filament, this process requires much time. An integral melt spinning process is being developed. The processes and their special features will now be described.

10.8.1 Sheet-type plastic optical fiber

A sheet-type plastic optical fiber is used as an optical line sensor and image guide for reading drawings and detecting defects. It is manufactured as follows.

The core and sheath materials are spun through a composite spinning nozzle in the same way as the single filament plastic optical fiber except that the spinning nozzle has many circularly arranged orifices as illustrated in Fig. 10.13. Many fibers are brought near to each other along the spin-guide and are stuck together in the form of a circular arc on a sticking guide located slightly apart from, and directly below, the spin-guide. They are then passed through an arranging guide located below the sticking guide and the fibers arranged in a straight line are drawn off by the nip rolls. Since the transit distances of the fibers from the spinning nozzle to the spin-guide are equal, uniform sheet type optical transmitters can be manufactured without deformation by processing.[19]

10.8.2 Block type plastic optical fiber

An image fiber composed of a bundle of many ultra-fine optical fibers has been used for endoscopes and the like. Since multi-component glass and quartz are brittle, further development of plastic image fibers is expected. In 1988 Mitsubishi Rayon developed and commercialized a plastic image fiber composed of a bundle of approximately 1500 step-index type plastic optical fibers 10–20 μm in diameter based on their

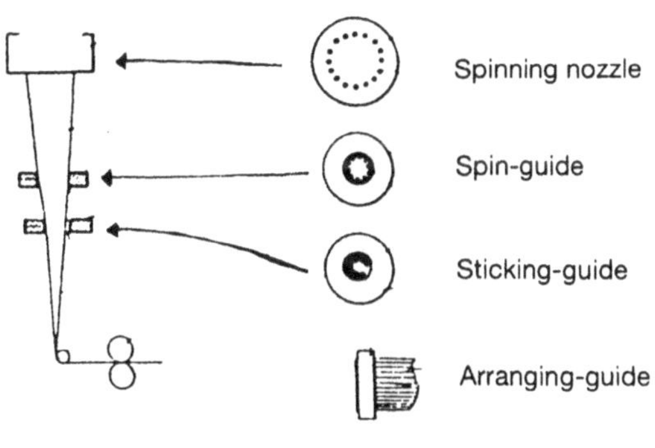

10.13 Process for manufacturing sheet type POF.

precise composite melt spinning technique. Following further programs for thinning the fiber and improving the resolving power, a plastic image fiber composed of a bundle of approximately 3000 ultrafine optical fibers 10 μm or less in diameter was developed in April 1991. A sectional photograph of it is shown in Fig. 10.14. The figure reveals that the circular plastic image fiber with a diameter of approximately 0.5 mm has approximately 3000 plastic optical fibers very closely packed. Each fiber corresponds to one picture element and is approximately 9 μm in diameter. The diameter of core and the thickness of sheath are approximately 7 μm and 1 μm, respectively. The plastic image fiber is characterized by higher flexibility, greater flexing resistance and resolving power and a brighter image than those of a glass image fiber. The modulus of elasticitiy of the plastic image fiber is 1/10 or less that of a quartz image fiber and it is very flexible. The brightness of image is expressed by:

$$E = F K_c \qquad [10.1]$$

where E = brightness of image, F = performance index of the plastic optical fiber constituting a plastic image fiber and K_c = ratio of core area to the total cross-sectional area of fiber. Using fibers with the same K_c, E is proportional to F. F is expressed by the following formula:

$$F = (\mathrm{NA})^2\, 10^{(-\alpha l/10)} \qquad [10.2]$$

where NA = number of apertures, α = transmission loss and l = length of fiber.

Although the plastic image fiber has a number of apertures as large as 0.5 and the transmission loss is as high as 600 dB/km, the plastic image fibers that are only a few metres long have larger F values than those of image fibers composed of other materials, as shown in Fig. 10.15.[20]

The spinning nozzle used for plastic image fibers is illustrated in Fig. 10.16. The molten core material fed to the nozzle is divided into core-forming plates and the sheath and intervening materials are fed to the circumference of the core inside the nozzle. A fiber composed of concentrically arranged core–sheath intervening components is discharged from the outlet and integrated into a uniformly arranged plastic image fiber at the integrating nozzle.[21]

Since organic optical fibers have special features for short-distance optical transmission media including low transmission loss and high heat resistance, various applications are expected. It is important to develop more reliable products as well as to reduce the transmission loss and develop wide range fibers for extending the present applications. It is also necessary to actively develop systems including data links and sensors as well as to develop the fiber itself.

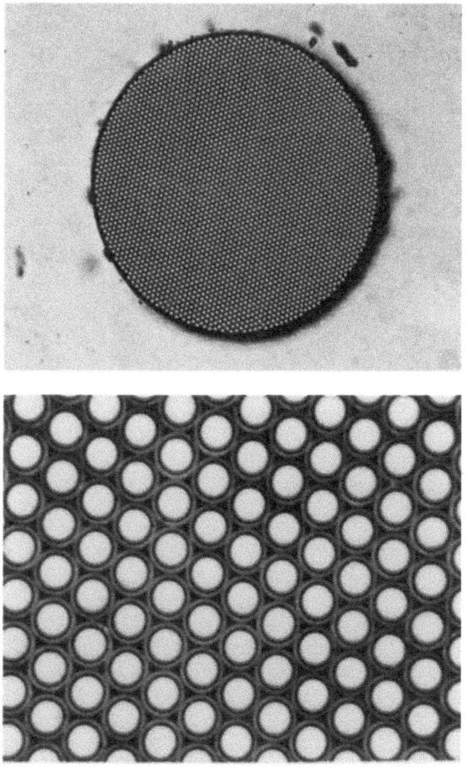

10.14 Cross-section of plastic image fiber.

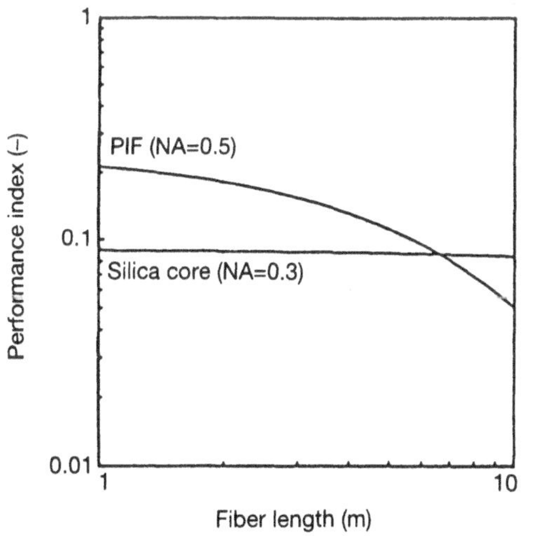

10.15 Relationship between fiber length and performance index.

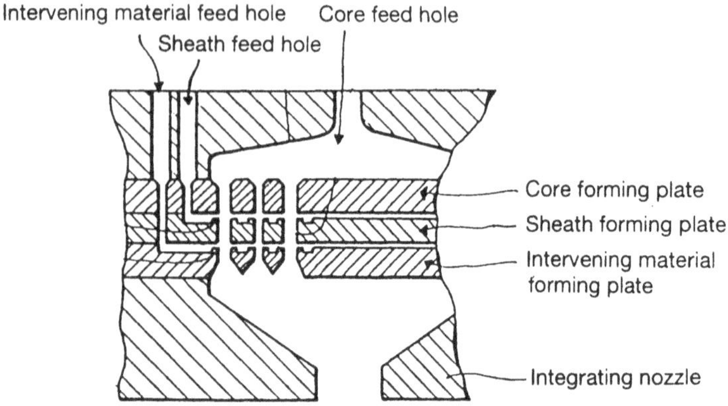

10.16 Spinning nozzle for plastic fiber.

References

1 Kaino T, *Kobunshi Ronbunshu*, **46**, 257, 1985.
2 Otsuka Y, Koike Y and Nihei E, *Polym. Prep. Jap.*, **39**, 3423, 1990.
3 Koike Y, Inai A, Isei H, and Nihei N, *Polym. Prep. Jap.*, **40**, 500, 1991.
4 Jap. Pat. (Laid open), 83-193,502, 86-176,902.
5 Jap. Pat. (Laid open), 82-81,303, 82-96,303
6 Jap. Pat. (Laid open), 82-104,906, 83-118,603, 85-220,303, 88-175,805.
7 Jap. Pat. (Laid open), 83-88,701, 88-94,203.
8 Jap. Pat. (Laid open), 83-132,028, 83-132,533, 84-133,206.
9 Jap. Pat. (Laid open), 84-328,168.
10 Miyashita T, *Kenkyu Jitsuyoka Hokoku*, **22**, 2467, 1973.
11 Jap. Pat. (Laid open), 88-192,976
12 Maclean D L, *Trans. Soc. Rheology*, **17**, 385, 1973.
13 White J L and Lee B L, *Polym. Eng. & Sci.*, **15**, 481, 1975.
14 Jap. Pat. (Laid open), 86-262,706.
15 Jap. Pat. (Laid open), 85-42,712.
16 Jap. Pat. (Laid open), 86-259,202.
17 Jap. Pat. (Laid open), 86-259,203.
18 Jap. Pat. (Laid open), 86-262,707.
19 Jap. Pat. (Laid open), 86-117,202.
20 Suzuki F, *Sen-i Gakkaishi*, **47**, 62, 1991.
21 Jap. Pat. (Laid open), 87-3,038.

Appendix
Microscopic views of Shingosen

Shingosen are advanced synthetic fibers and fabrics, usually the product of combination of advanced technologies. They have opened up a new field of high-quality textiles. The best way of conveying the nature of these products and their characteristics, short of handling them, is to provide micrographs taken by an electron microscope. This appendix contains photographs of SEM images of Shingosen currently produced in Japan, with a brief summary of their commercial names, their characteristic properties, the technologies involved in producing them, and the fields of application.

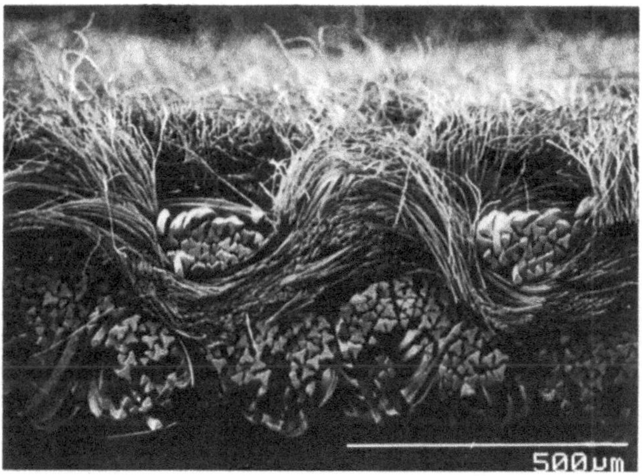

UTS (Toray)

Main technology: Finely crimped 0.06 denier (2 μm diameter) filaments. A-Z random processing (large interfilamental space). Mixed filaments.

Examples of final products: Jackets, suits, and coats.

Characteristics: Powdery smoothness in touch, warm and light.

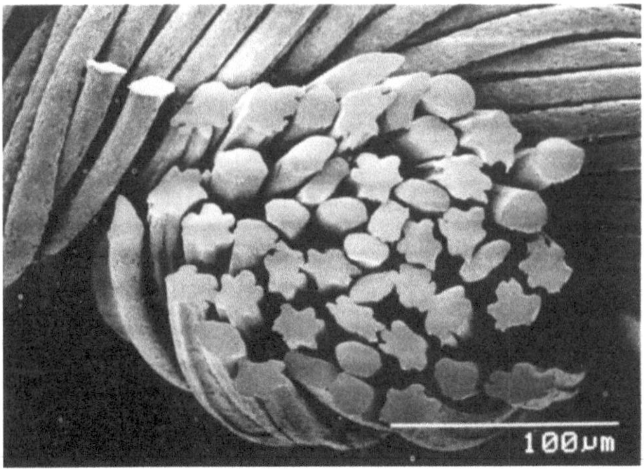

CEO α (Toray)

Main technology: Minute gaps between filaments. Various fiber cross-sections. Mixed filaments.

Examples of final products: Blouses and summer suits.

Characteristics: Good moisture absorbency, dry touch.

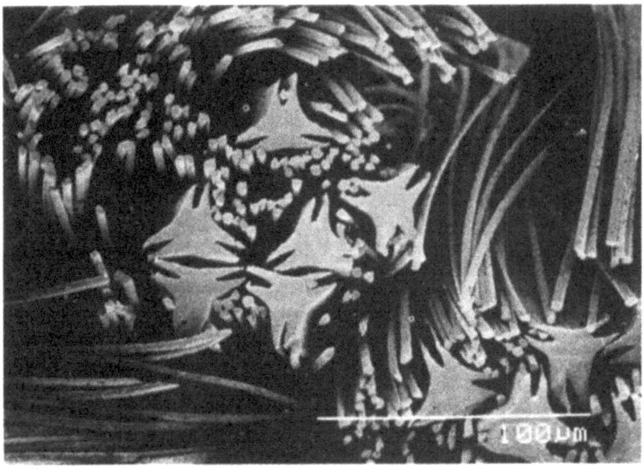

SILLOOK TIFFARA (Toray)

Main technology: 0.2 denier filaments containing ceramics. Tri-petal cross-section.

Examples of final products: Blouses, dresses and jackets.

Characteristics: Dry and cool hand, bulk and resilient, silky luster, good tailoring.

RIRANCHE (Toray)
Main technology: Multi-entangled composite yarn texturizing.
Examples of final products: Ladies blouses and jackets. Men's jackets.
Characteristics: Lustrous appearance, soft, bulky, good drape, cool touch.

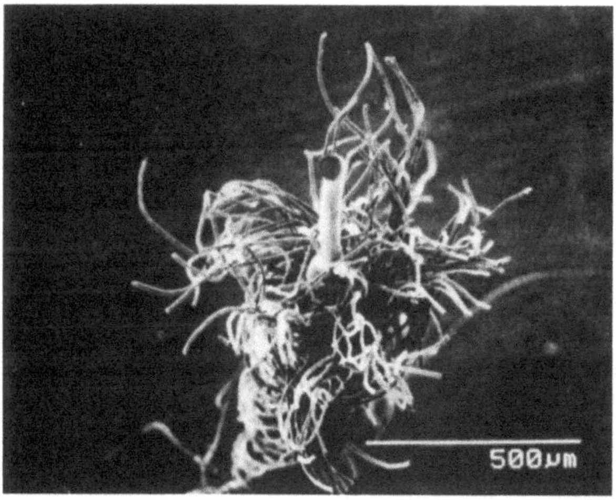

DUARA (Toray)
Main technology: Staple and filament composite spun yarn.
Examples of final products: Dresses, Formal suits.
Characteristics: Dry touch, bulky, good dyeability.

PICEME (Toray)

Main technology: Polyester/polyamide conjugate spinning. Filament-splitting during textile processing.

Examples of final products: Blousons. Skiwear.

Characteristics: Touch like rose petal, deep color and luster, water repelling, waterproof and vapor permeable.

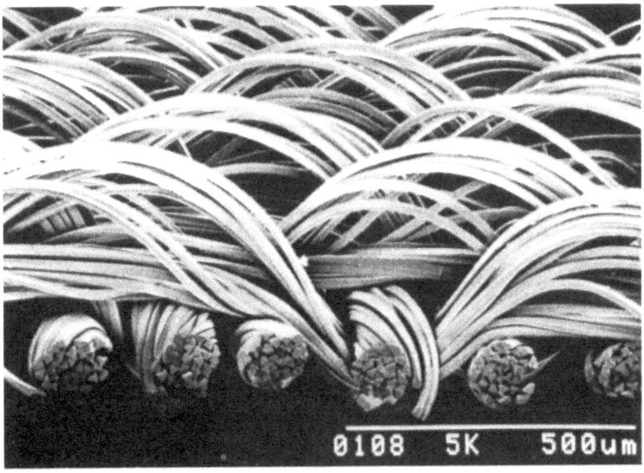

SILLOOK SILDEW (Toray)

Main technology: Multi-stage differential shrinking.

Examples of final products: Blouses. One-piece dresses.

Characteristics: Supple, silk-like, good bulk, mild luster, deep color.

CONCLAIRE (Toray)

Main technology: Uneven filaments of irregular denier. Different shrinkage in each filament. Loop-type randomly entangled filaments.

Examples of final products: Vests. Blouses.

Characteristics: Good tailored appearance, light and warm.

SILLOOK ROYAL S (Toray)

Main technology: Latent differential shrinkage of filaments. Multi-stage shrinking. Triangular cross-section with slits at apexes.

Examples of final products: Western style dresses. Kimonos.

Characteristics: Silky, bulky and resilient feel, silk-like scroop, good drape and liveliness.

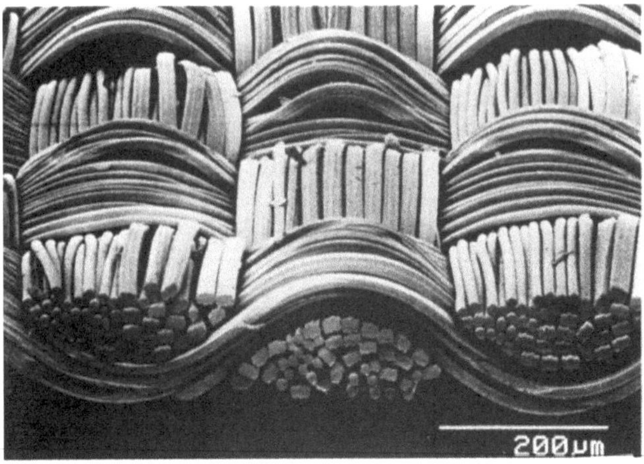

CEO (Toray)

Main technology: Submicron-controlled non-circular cross-section. Longitudinal micro-uneveness.

Examples of final products: Blouses. Jackets.

Characteristics: Cool and dry in touch, hemp-like hand, easy tailoring.

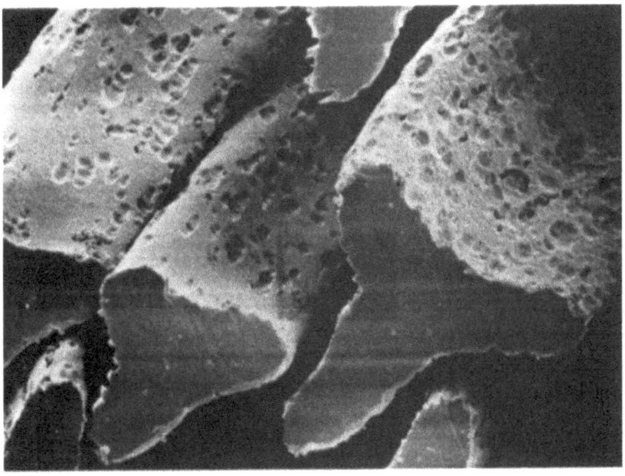

SILLOOK CHATELAINE (Toray)

Main technology: Controlled micro-craters. Latent multi-stage shrinking. Trefoil cross-section.

Examples of final products: Jackets. Skirts.

Characteristics: Silk-like touch, characteristic body.

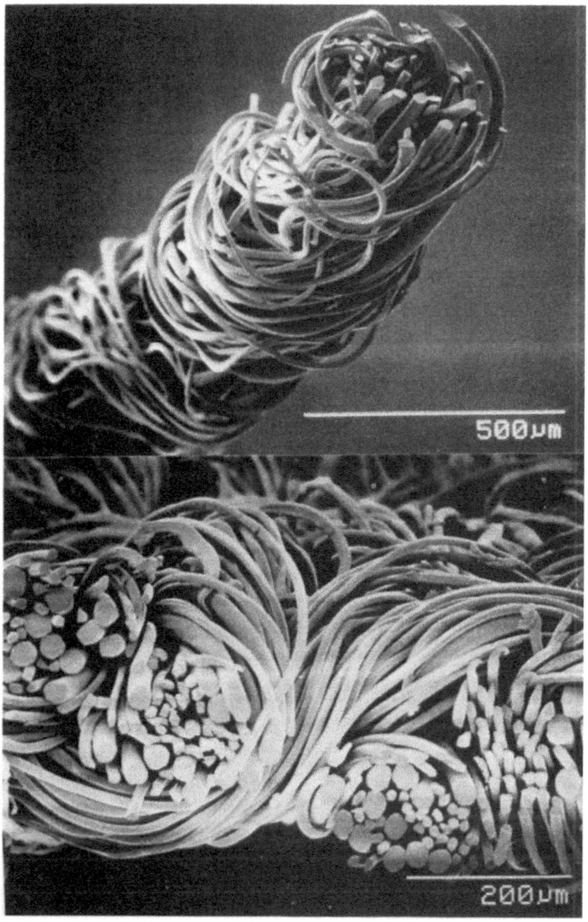

MALOR (Toray)
Main technology: Side-by-side type conjugate yarn. Uneven filament deniers. Random degree of crimp.
Examples of final products: Suits and jackets for men. Ladies' suits.
Characteristics: Elastic, easy tailoring, light and bulky, supple feel, resilience and drapeability.

AJENTY (Teijin)

Main technology: Woven fabric with microwaves on surface due to differential elasticity of trilobal filaments.

Examples of final products: Ladies' wear. Men's wear.

Characteristics: High bulk and softness with fine-brushed surface effect, clear and deep color.

GUARDIA (Teijin)

Main technology: Uneven drawing of bicomponent yarn.

Examples of final products: Ladies' wear.

Characteristics: Natural appearance through uneven (deep-colored thin and pale-colored thick) pattern.

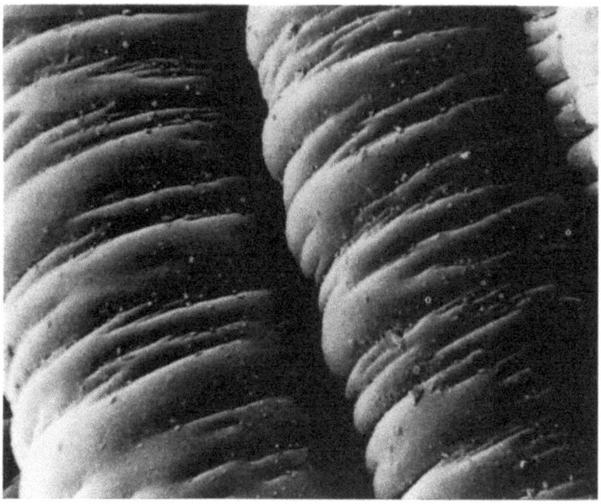

TEPLA (Teijin)

Main technology: Controlled drawing to yield filament fineness varying randomly along fiber axis. Scaly unevenness on filament surface.

Examples of final products: Ladies' wear. Men's wear.

Characteristics: Spun-like surface feel similar to natural spun silk, natural appearance.

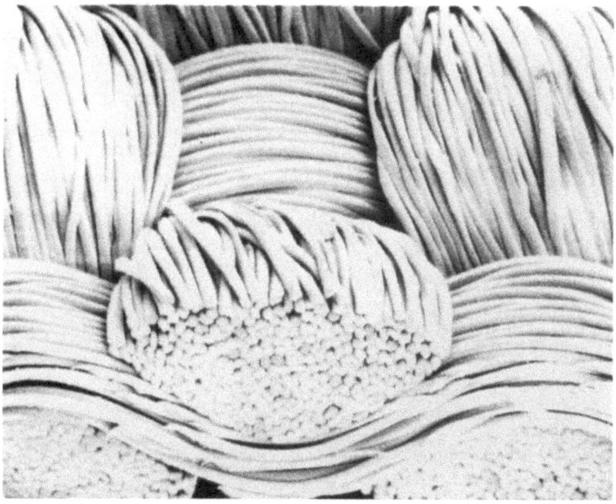

CONDENIER (Teijin)

Main technology: High-density fabric, woven from micro-fiber yarn and relaxed at finishing.

Examples of final products: Sports wear. Coats.

Characteristics: Waterproof cloth of uncoated type without resin-coating or film lamination, high moisture vapor permeability.

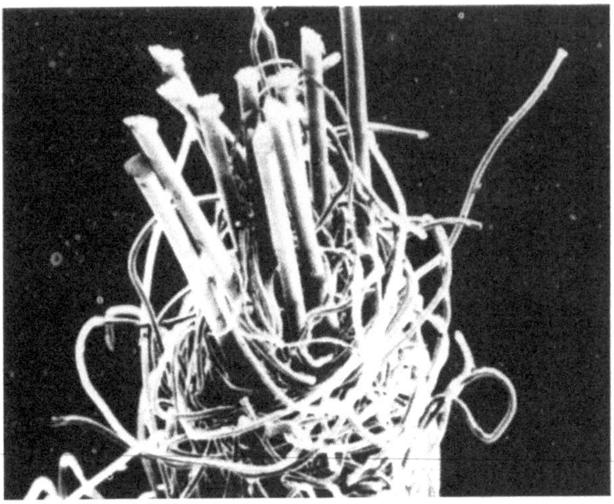

XOXO (Teijin)

Main technology: Special spinning with thick filaments as a core covered by fine short fibers made by tearing tows (*Microfuzz* wrapped structure).

Examples of final products: Ladies' wear. Sports wear.

Characteristics: Spun-like touch and high resilience.

XOXO-L (Teijin)

Main technology: Basically the same as XOXO, but finer denier filaments in core.

Examples of final products: Ladies' wear.

Characteristics: Linen-like appearance, moderate stiffness, peculiar drapeability.

SILDOLL LEGERETE (Teijin)

Main technology: Special spinning 'micro-random spinning'. Techniques for fibillar structure.

Examples of final products: Ladies' wear.

Characteristics: Natural appearance, similar to sanded silks, scroop effect.

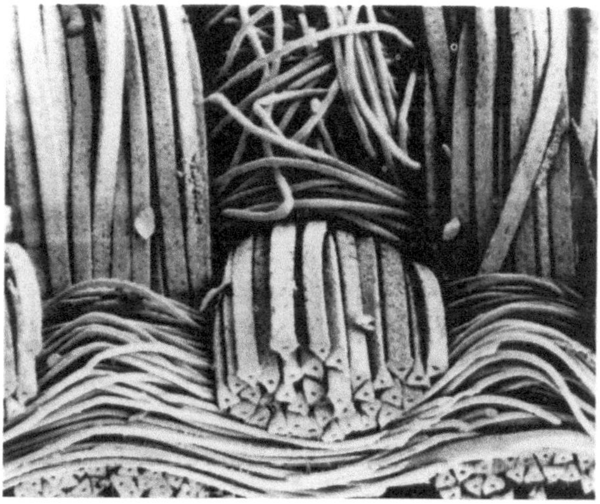

WELLKEY-X (Teijin)

Main technology: Combination of WELLKEY yarn made of hollow fibers with varying shrinkage (as warp yarn) and XOXO yarn of microfuzz wrapped structure (as weft yarn).

Examples of final products: Ladies' shirts and blouses. Men's shirts.

Characteristics: Water-absorbing property, silk-spun-like feel, scroop effect.

ROCHET (Teijin)

Main technology: Bicomponent yarn without splitting after dyeing. Composed of mixed filaments of different deniers and shrinkage.

Examples of final products: Ladies' wear. Men's wear.

Characteristics: Silk-worsted handle, high stiffness and resilience.

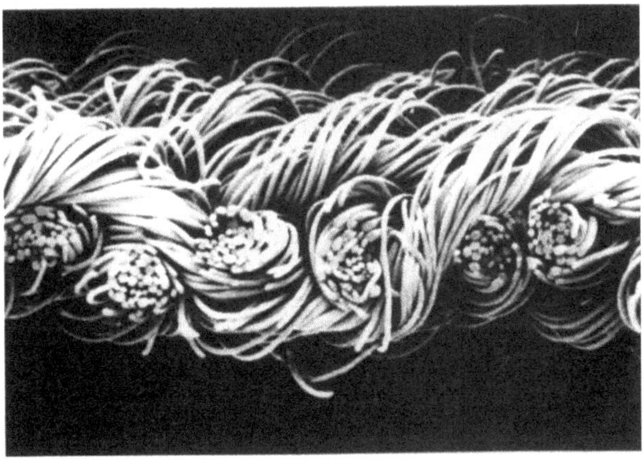

CONSOFF (Teijin)

Main technology: Air-texturing of bicomponent fibers. Yarn surface micro-fibers of various degrees of chain orientation and yarn center coarse filaments of high shrinkage.

Examples of final products: Men's wear. Ladies' wear.

Characteristics: High bulk and resilience, deep color, fine worsted handle.

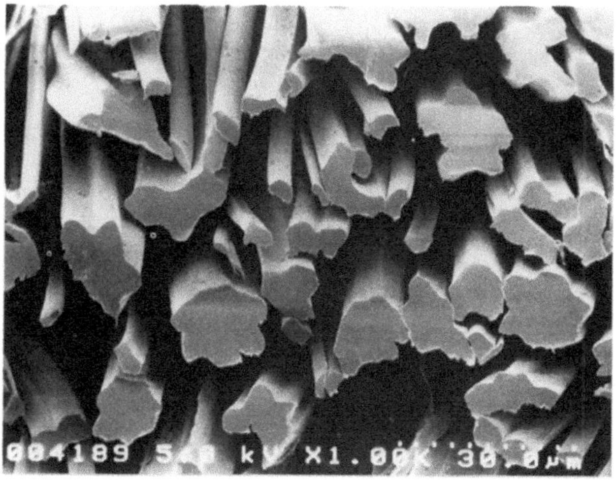

NACLE (Kanebo)
Main technology: Random cross-section conjugate yarn. Special finishing.
Examples of final products: Blouses. Dresses.
Characteristics: Spun-silk-like handle, dry touch, moderate resilency.

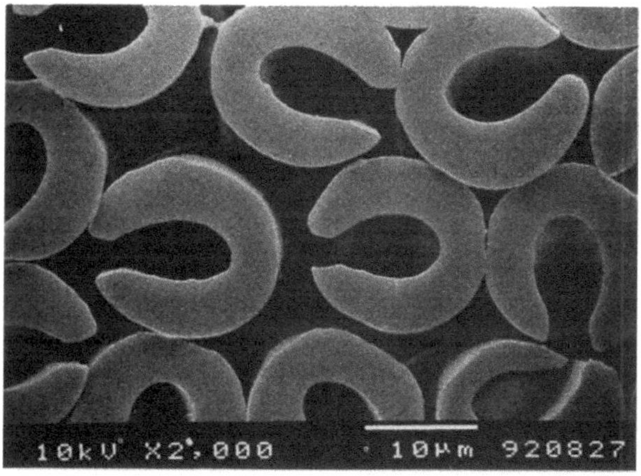

KILATT (Kanebo)
Main technology: C-shaped cross-section large-hollow fiber. Conjugate yarn. Texturizing possible.
Examples of final products: Sports wear. Casual wear.
Characteristics: Light, warmth-retentive, bouncy touch.

NAZCA (Kanebo)

Main technology: Micro-fiber from conjugate yarn. Differential shrinkage. Light brushing.

Examples of final products: Ladies' dresses. Blouses.

Characteristics: Moist touch, shadowy appearance, soft and bulky.

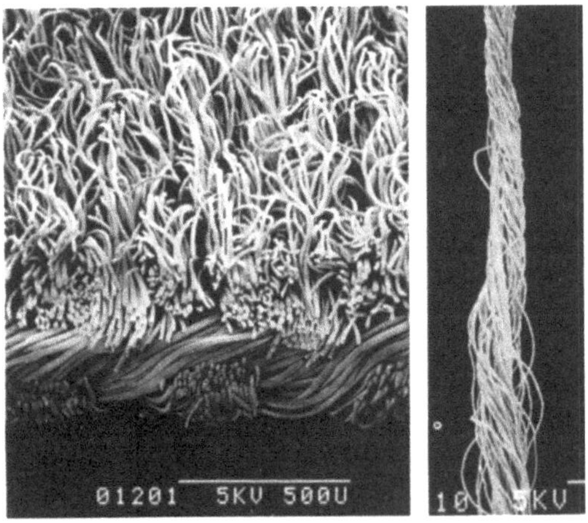

CASHMEENA (Kanebo)

Main technology: Composite textured yarn. Differential shrinkage. No napping/brushing.

Examples of final products: Ladies' suits. Blouses.

Characteristics: Cashmere-like, soft and bulky, superior drape.

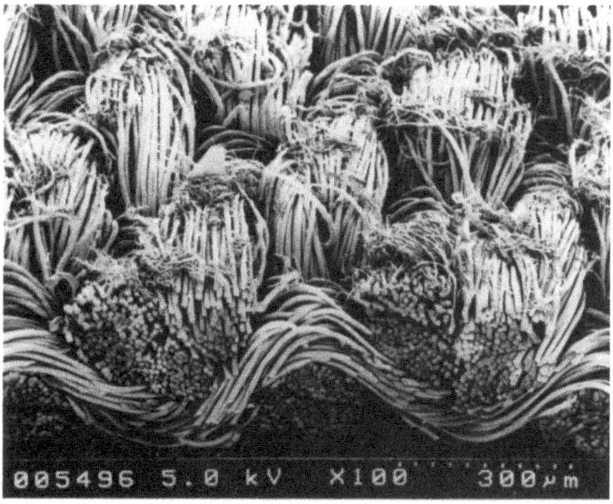

AOAI (Kanebo)

Main technology: Hybrid fabric of polynosic and shingosen. Fibrillation of polynosic by biological treatment.

Examples of final products: Blouses. Jackets.

Characteristics: Sophisticated shadowy appearance, soft and wrinkle-free, good dimensional stability and recovery.

OBJET (Kanebo)

Main technology: Composite textured yarn. Color mix. Differential shrinkage.

Examples of final products: Ladies' suits and bottoms.

Characteristics: Soft, tweed-like, stereographic surface, good airy drape.

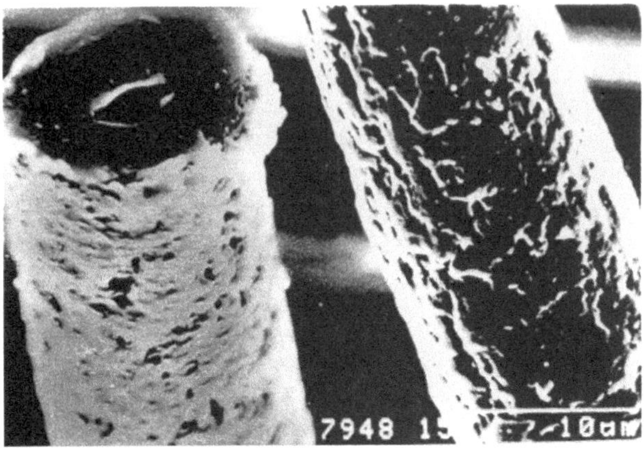

LOUVRO (Toyobo)

Main technology: Dispersion of fine ceramics which possess the same refraction as PET. Hollow fiber.

Examples of final products: Dresses. Blouses.

Characteristics: Lively resilience, dry touch, full dull luster.

ROSALLY (Toyobo)

Main technology: Special texturing. Mixed filaments of coarse deniers.

Examples of final products: Suits. Pants.

Characteristics: Soft and dry powdery touch, characteristic surface appearance, moderate resilience and bulk.

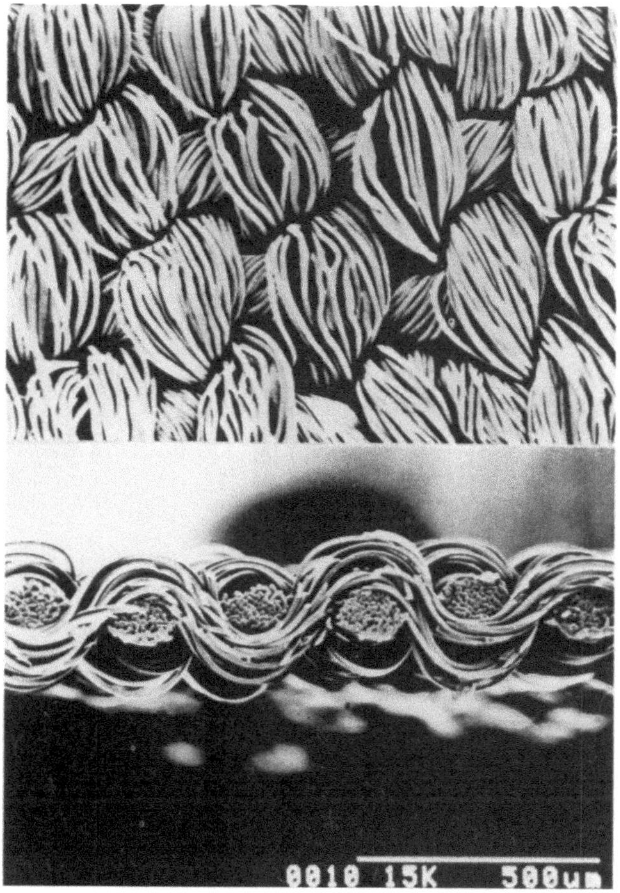

GEENA (Toyobo)

Main technology: Shape memory effect by molecular alignment technique at finishing. Mixed filaments. Special fabrication and finishing.

Examples of final products: Dresses. Blouses.

Characteristics: Soft, warm hand, deep color.

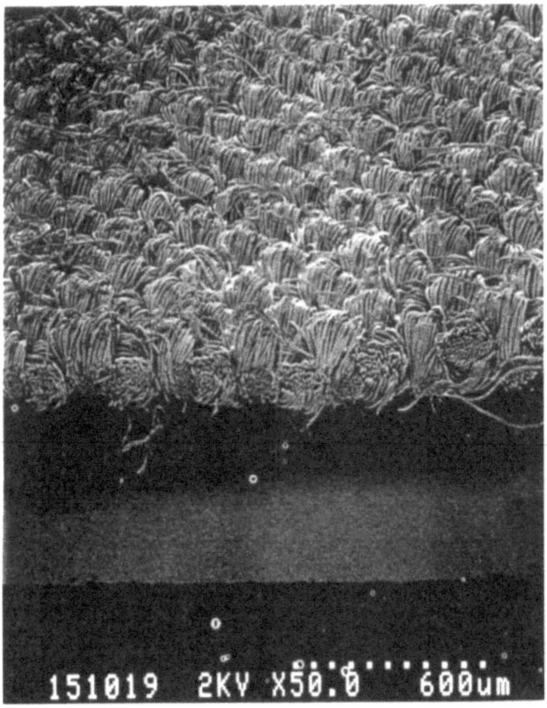

SOCIO (Toyobo)

Main technology: Uniformly mixed composite spun yarn composed of Shingosen and staple fiber.

Examples of final products: Suits. Coats.

Characteristics: Soft and bulky, good drape, powdery feel.

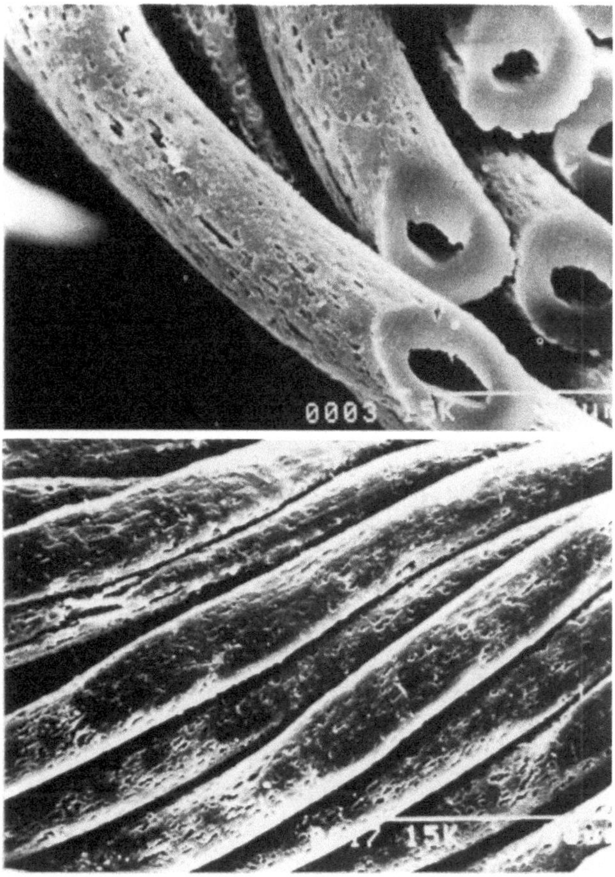

SHANDEL (Toyobo)

Main technology: Uniformly dispersed fine ceramics. Thick and thin yarn (controlled irregularity).

Examples of final products: Dresses. Blouses.

Characteristics: Lively resilience with good drape, natural unevenness, light.

BONANZA (Mitsubishi Rayon)

Main technology: Multi-layered composite yarn. Copolymer. Controlled molecular orientation (crystallization).

Examples of final products: Dresses. Jackets.

Characteristics: Wool-like-touch, high bulk, smoothness.

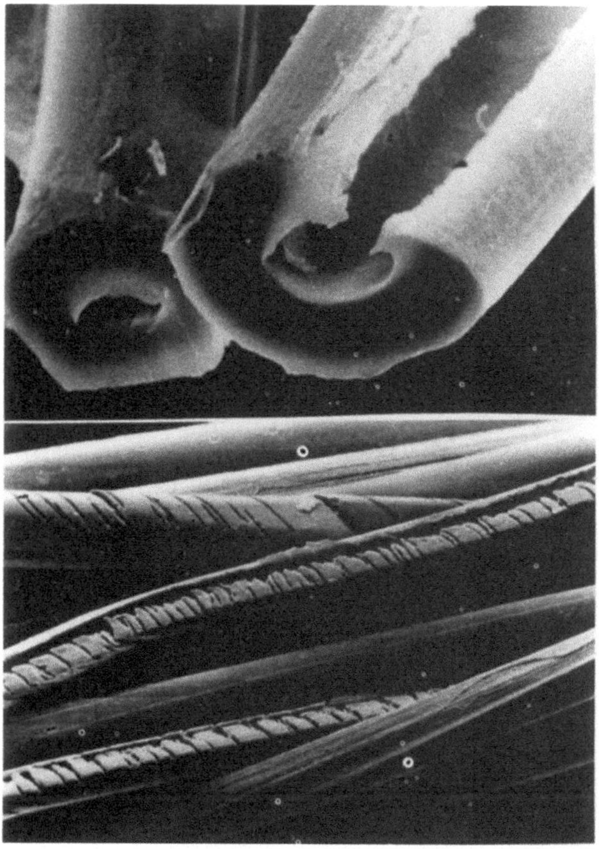

AEGE (Mitsubishi Rayon)
Main technology: Conjugate hollow filaments.
Examples of final products: Blouses.
Characteristics: High perspiration absorption, quick drying.

CRISETA (Mitsubishi Rayon)

Main technology: Micro-diversified thick and thin yarn. Special drawing and twisting.

Examples of final products: Dresses. Blouses.

Characteristics: Dry spun-silk hand, soft drape.

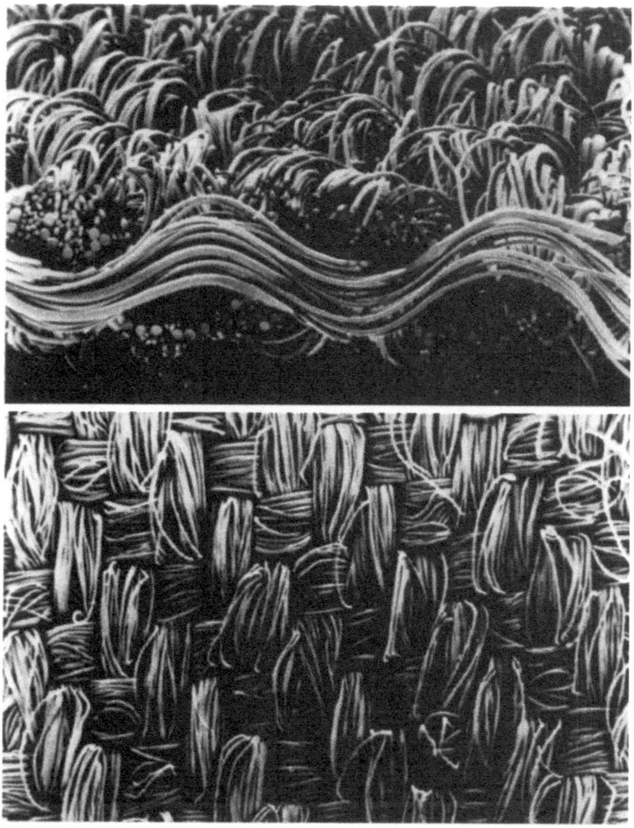

NAIVA (Unitika)
Main technology: Bicomponent fiber of nylon and EVAL. Sheath-core.
Examples of final products: Sports wear. Coats.
Characteristics: Peach-like type with high shrinkage.

FSY (Unitika)

Main technology: Bicomponent fiber consisting of two polyesters with different alkali-solubility. Separation by alkali-treatment.

Examples of final products: Ladies' coats. Wiping cloths.

Characteristics: Polyester microfiber of 0.1 dtex with wedge-shaped cross-section.

CLEMENT (Unitika)

Main technology: Modified air-texturing of two-layered yarn consisting of two polyesters. Modification of crystalline structure.

Examples of final products: Ladies' suits. Men's pants.

Characteristics: Worsted-like polyester fiber with deep dyeability.

ROMIEL (Kuraray)
Main technology: Super-micro staple fiber of 0.4 denier. Sheath-core spun yarn. Lightly brushed.
Examples of final products: Coats. Jackets.
Characteristics: Soft and smooth touch, uniform surface, high drapeability.

F F (Kuraray)

Main technology: Conjugate spinning. Fibrillation and splitting of fibers. Lightly brushed.

Examples of final products: Dresses. Coats.

Characteristics: Soft surface appearance, feather-like touch, moderate bulk.

DUO II (Asahi Kasei)

Main technology: Low-shrinkage high-speed-spun PET yarn mixed with high-shrinkage conventional PET yarn.

Examples of final products: Blouses. Coats.

Characteristics: Soft and dry touch, light, silk-like luster.

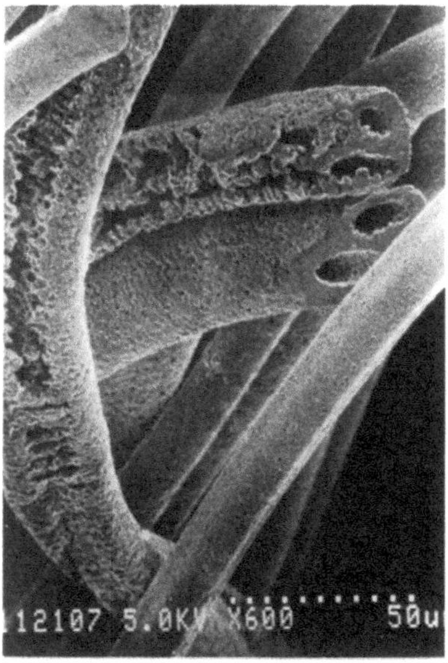

GULKA (Asahi Kasei)

Main technology: Triangular cross-section with three hollows. Micro-craters. Thick and thin yarn.

Examples of final products: Blouses, suits.

Characteristics: Moderately bulky hand, dry touch, shades against ultraviolet rays

Index

Subject index

Company index

Page numbers referring to patents are italicized

Trade name index

FSC
www.fsc.org
MIX
Papier | Fördert
gute Waldnutzung
FSC® C083411

Kolibri 360 GmbH
Ferdinand-Jühlke-Straße 7
99095 Erfurt, Deutschland
produktsicherheit@kolibri360.de